Comprehensive Quality by Design in Pharmaceutical Development and Manufacture

AA003451

Topical Conference at the 2010 AIChE Annual Meeting

Salt Lake City, Utah, USA
7-12 November 2010

ISBN: 978-1-61782-146-2

Printed from e-media with permission by:

Curran Associates, Inc.
57 Morehouse Lane
Red Hook, NY 12571

Some format issues inherent in the e-media version may also appear in this print version.

Printed by Curran Associates, Inc. (2010)

For permission requests, please contact AIChE
at the address below.

AIChE
3 Park Avenue
New York, NY 10016-5991

Phone: (203) 702-7660
Fax: (203) 775-5177

www.aiche.org

Additional copies of this publication are available from:

Curran Associates, Inc.
57 Morehouse Lane
Red Hook, NY 12571 USA
Phone: 845-758-0400
Fax: 845-758-2634
Email: curran@proceedings.com
Web: www.proceedings.com

TABLE OF CONTENTS

(51a) Scale Independent Approach for Drug Product Development by High Shear Granulation Process

Gavin K. Reynolds, Bindhu Gururajan, Jonathan Sutch and James Kraunsoe, Pharmaceutical Development, AstraZeneca, Macclesfield, United Kingdom

Pharmaceutical drug product development is driven through empirical experimentation. Factorial experimental design methodologies are utilized to improve the efficiency of this approach. A key unit operation in the manufacture of many immediate release tablets is high shear wet granulation. High shear wet granulation is a time-dependent process with many simultaneous physical mechanisms occurring and therefore development work is often focused on the impact of this unit operation on final product quality attributes.

During process development, experimental designs will include typical factors such as impeller speed (rpm) or spray rate (g/min), which tend to be scale-dependent. Analysis of such an experimental design provides important process knowledge and understanding and often includes empirical models relating process parameters to product properties. However, as the input factors are scale-dependent, it remains unclear how to quantitatively transfer the empirical models and understanding to different equipment scales.

Historically, scale-independent parameters such as tip speed and Froude number have been developed for the high shear granulation process, many based on the principles of dimensionless numbers. These scale-independent parameters are typically focused on different mechanisms of the wet granulation process and therefore can provide conflicting scale-up relationships. The challenge remains in terms of how to utilize factorial experimental design data across manufacturing scales, and how to identify the most appropriate scale-up relationships.

A methodology is presented whereby experimental data is analysed in the context of a range of scale-independent parameters from the literature. Statistical techniques are used to identify the sub-set of parameters that provide the best description of the product attributes. These scale-independent parameters are then used as a basis for developing a scale-independent understanding of the manufacturing of a particular formulation.

A case study is presented where this methodology was retrospectively applied to a product that had been manufactured over scales ranging from 2.5 to 180 kg. Throughout development, the method of water addition had been qualitatively identified as important for the product intermediate quality. Based on this observation, the dimensionless spray flux was included as part of retrospective analysis using scale independent parameters. The dimensionless spray flux definition of Litster et al. (Pow. Tech. 114, pp32-39, 2001) is based on carefully controlled experimental studies examining droplet spray on the exposed powder surface of a spinning riffler. The approach has been extended in order to provide a consistent definition that only depends on easily measurable and controllable wet granulation process parameters. The key process parameters that are included in the definition are shown in Figure 1.

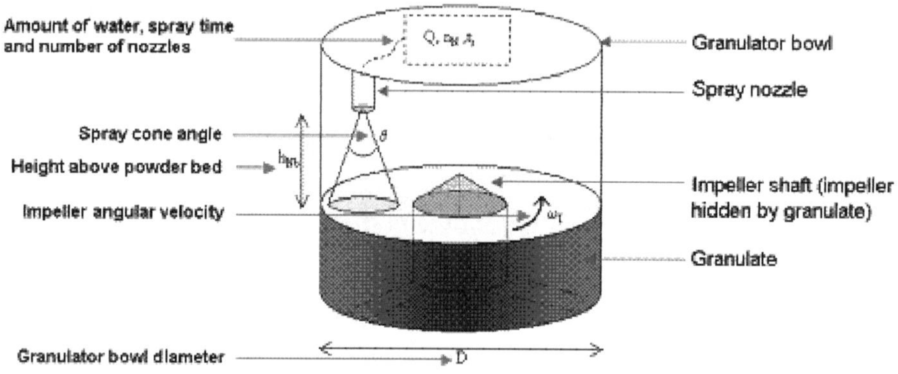

Figure 1 Key parameters included in dimensionless spray flux calculation

Retrospective analysis identified that the scale-independent parameter of dimensionless spray flux had been maintained within a consistent range of 0.2 to 1.4 leading to consistent intermediate granule attributes and end product quality. It is not possible to directly compare the numeric range with the regimes proposed by Litster et al. (Pow. Tech. 114, pp32-39, 2001) due to the changes in the method of calculation. Nevertheless, the approach appears relevant in this case, whereby the majority of product experience has been generated within a relatively narrow regime and resulted in controlled granule growth and consistent product quality attributes. The extension of the dimensionless spray flux calculation therefore allows future changes in granulation scale and process parameters to be constrained multivariately, such that the high shear wet granulation process will remain in a spray controlled regime consistent with demonstrated experience and assuring product quality attributes.

See more of this Session: Scale-up of Pharmaceutical Manufacturing Processes: Toward a QbD Approach

See more of this Group/Topical: Topical i : Comprehensive Quality by Design in Pharmaceutical Development and Manufacture

(51b) A Quality by Design Approach to Scale-up of the Melt-Spray-Congeal Multiparticulate Process

Andrew Prpich, Pharmaceutical Development, Pfizer Global Research & Development, Groton, CT, Pankaj Doshi, Chemical Engineering and Process Division, National Chemical Laboratory, Pune, India and Louis Topper, Department of Mechanical Engineering, University of California, San Diego, San Diego, CA

The melt-spray-congeal (MSC) process is a rotary atomization technique for the production of a multiparticulate pharmaceutical dosage form. The final drug product consists of a wax matrix microsphere with the active pharmaceutical ingredient (API) dispersed throughout. The drug is released from the microsphere core via porous diffusion through either channels formed as the API dissolves or pores created by various soluble excipients that can be added to the formulation. The rate of dissolution of the drug is typically correlated to the particle size of the microsphere. Thus, it is imperative be able to control the microsphere size during manufacturing to attain a reproducible release profile. For the MSC process the key process parameters which influence the microsphere size are the speed of the spinning disk (rpm), the viscosity of the drug-melt suspension, and the operating temperature of the system.

The main objectives of this work were: (1) to investigate how these process parameters influence the key quality attributes (microsphere size and dissolution) of the final drug product, and (2) to develop predictive models for the melt rheology and microsphere size. To achieve these objectives a full 2^3 factorial design of experiment (DOE) was run using a model drug system. The factors that were investigated in the DOE included % drug loading (correlated to melt viscosity), disk speed (rpm), and operating temperature. In general it was found that increasing API loading (correlated to higher melt viscosity), decreasing temperature, and decreasing disk speed resulted in increased microsphere particle size. In terms of performance it was found that above a certain threshold of drug loading the dissolution profile was insensitive to microsphere particle size. The melt viscosity data was used to develop a power law model which describes the shear dependent viscosity behavior of the drug-melt system. In addition, a predictive model was developed to determine the mean microsphere diameter as a function of melt viscosity and disk speed. A comparison of measured and predicted mean microsphere size for 14 different MSC runs show that all model predictions are within 10% of the experimental data. These two models tie the formulation and the MSC operating conditions to the resulting microsphere particle size and can be used to reduce the time and number of experiments required to develop a melt-spray-congeal formulation.

See more of this Session: Scale-up of Pharmaceutical Manufacturing Processes: Toward a QbD Approach
See more of this Group/Topical: Topical i : Comprehensive Quality by Design in Pharmaceutical

(51c) Scale-up of High Shear Wet Granulation of Active Formulations Using a Dimensional Analysis Approach

Intan Munirah Hamdan, Chemical Engineering, Purdue University, West Lafayette, IN and **William R. Ketterhagen**, Pfizer Inc., Groton, CT

Wet granulation is a key process in the pharmaceutical industry in which a liquid binder is distributed into a powder blend to achieve better powder blend characteristics such as flowability and content uniformity. The wet granulation parameters required to obtain the desirable powder characteristics are first determined in small scale and are then scaled for manufacturing purposes. Since the wet granulation process is inherently complex, a significant amount of time is spent on adapting the process for larger scales. It is thus desirable to shorten this scaling-up procedure by using the knowledge obtained at smaller scales and developing dimensionless relationships that can then predict the behavior at larger scales with minimal or no adjustments. In our study, we use the scale-up methodology based on the dimensional analysis model developed by Anne Faure et al. (Faure, A., et al (1999) "Applicability of a scale-up methodology for wet granulation processes in Collette Gral high shear mixer-granulators", European Journal of Pharmaceutical Sciences, 8 (1999), 85-93), in which the coefficients 'a' and 'b' of the model characterize the scaling model for that particular formulation. It is assumed that the net power consumption is a good indicator of the stage in the granulation process, and hence is used to predict the wet granulation end point.

Wet Granulation Scale-Up Model: $\log Np = a \cdot \log(\Psi Re \cdot Fr \cdot \text{fill ratio}) + b$

Np (Power number), ΨRe (Pseudo-Reynolds number), Fr (Froude number), fill ratio are the dimensionless groups derived from the parameters that govern the wet granulation process.

Experiments were conducted in a 10-L Collette Gral to develop scale-up correlations for five formulations – two placebo formulations, and three active formulations. It was found that the model was a good representation of the wet granulation process for all five formulations. The slopes of the active formulations were not statistically different from one another, but were different from the placebo formulations (which shared similar slopes). The intercepts were statistically distinct for all formulations except for two active formulations. The differences in model coefficients for the different formulations reaffirm the belief that wet granulation parameters are formulation specific and that scale-up correlations are also formulation specific. Additionally, these differences could be an indication that the effects of factors affecting wet granulation (i.e. powder particle size distribution) manifest themselves in the dimensionless groups. Experiments were then conducted to validate the correlations: the net power consumption needed to attain the granulation end point for a placebo formulation run at an impeller speed of 450rpm was calculated from the model, and used as an end point. The granule properties of the validation experiments were found to be in spec. Finally, experiments were conducted in a 25-L Collette Gral. After accounting for the different surface area to volume ratios of the bowls, the experiments were found to be in good agreement with the model developed at the 10-L scale. Thus, once a model has been developed for a particular formulation and operating procedure, it can be used to predict the granulation end point across all scales.

See more of this Session: Scale-up of Pharmaceutical Manufacturing Processes: Toward a QbD Approach
See more of this Group/Topical: Topical i : Comprehensive Quality by Design in Pharmaceutical Development and Manufacture

(51d) A Regulators Perspective On the Evaluation and Control of Ingredient Quality

Steven Wolfgang, Office of Compliance, Division of Manufacturing and Product Quality, FDA Center for Drug Evaluation and Research, White Oak, MD

The compendial specifications are limited in their ability to define and quantify the quality of ingredients as it pertains to their functional performance in processes and products. The properties of materials will affect how they behave and the observed behavior can change dramatically as the scale of operation increases. A significant body of information about ingredient physicochemical properties gathered in early phase material characterization studies should be useful when entering the scale up phase of development. Better understanding of the interrelationships involving measurable ingredient properties can lead to better predictions of how the ingredients will behave as the process is scaled up, including how ingredient behavior will be influenced by process variables, equipment and environmental conditions. Manufacturers should find a body of knowledge and data pertaining to ingredients useful in their efforts to design more robust and predictable processes. Supplier management - including qualification, auditing and monitoring - will be discussed as tools that help build and maintain control over ingredient quality to promote reliable processing at scale-up and throughout the product lifecycle.

See more of this Session: Scale-up of Pharmaceutical Manufacturing Processes: Toward a QbD Approach
See more of this Group/Topical: Topical i : Comprehensive Quality by Design in Pharmaceutical Development and Manufacture

(51e) Toward a More Reliable USP Dissolution Testing Apparatus II

Yimin Wang, Otto H. York Department of Chemical, Biological and Pharmaceutical Engineering, New Jersey Institute of Technology, Newark, NJ and Piero Armenante, Chemical, Biological and Pharmaceutical Engineering, New Jersey Institute of Technology, Newark, NJ

In the pharmaceutical industry, dissolution testing is a critical step in quality control and a standard method for assessing batch-to-batch consistency of solid oral drug delivery systems, such as tablets. One of the most widely used dissolution test devices is the UPS Apparatus II (paddle). Even to this day, dissolution testing remains susceptible to significant error and test failures. Previous studies indicate that poor reproducibility of dissolution testing data and inconsistency of dissolution results can arise from the complex hydrodynamics present in the unbaffled, hemispherical-bottom, agitated vessel that constitute the UPS Apparatus II. In previous work, our group investigated the hydrodynamics in the standard UPS apparatus II and showed that the velocity distribution in the bottom region of the vessel, where the tablet is typically located and where dissolution occurs, is highly non-uniform and the flow pattern is highly variable.

In this work, a modified USP Apparatus II was constructed in which the impeller was placed 8 mm off center. Dissolution tests where calibrators tablets (Prednisone) were placed at different positions on the vessel bottom were conducted using both the standard USP Apparatus II and the modified apparatus. The dissolution profiles were compared with each other and the difference factors f_1 and the similarity factors f_2 were calculated from the dissolution profiles.

The dissolution profiles in the modified apparatus were not significantly affected by tablet location as confirmed by the value of the factors f_1 and f_2 ,which were well within the accepted ranges and did not change appreciably with the tablet location. By contrast, the corresponding dissolution tests in the standard apparatus failed these similarity tests. In addition, the flow fields near the vessel bottom were obtained via CFD simulation and were found to be significantly more uniform in the modified USP Apparatus II than in the standard apparatus.

The proposed modified apparatus has the potential of becoming a valid alternative to the standard USP dissolution testing apparatuses used for dissolution testing.

See more of this Session: Scale-up of Pharmaceutical Manufacturing Processes: Toward a QbD Approach
See more of this Group/Topical: Topical i : Comprehensive Quality by Design in Pharmaceutical Development and Manufacture

(51f) Using Qualitative Models to Capture and Demonstrate Process Understanding

Ian Houson, BRITEST Ltd., Cheshire, United Kingdom

This presentation will cover how qualitative models can be used to capture and demonstrate process understanding and improve communication between disciplines, across geographical locations, languages and cultures both within, and between, organisations and regulators within a QbD framework. The presentation will highlight how the use of pictorial, information rich tools can improve learning between disciplines in order to bring the fundamental scientific ideas and concepts together. It will also demonstrate how to significantly reduce risks during scale-up through improved process understanding by incorporating both physical and chemical aspects into the models. This presentation will explain how the generation of qualitative models assessing separate aspects of the process, allows for structured yet rapid identification of process improvement options and key data gaps. It will also identify likely critical process parameters (CPPs) and provides a sound scientific rationale for where DOEs should be placed within the design space. Use of these qualitative models will also identify where the targeted use of advanced modelling tools and techniques will add most value.

See more of this Session: Scale-up of Pharmaceutical Manufacturing Processes: Toward a QbD Approach
See more of this Group/Topical: Topical i : Comprehensive Quality by Design in Pharmaceutical Development and Manufacture

(51g) Characterization of Tablet Attritional Forces in a Friability Tester and Film Coating Pan Using DEM

Rahul Bharadwaj, William R. Ketterhagen and Bruno C. Hancock, Pfizer Inc., Groton, CT

The objective of this work was to better understand the forces generated on tablets in the friability tester and pan coater using discrete element method (DEM) simulations. Such systems are used in the pharmaceutical industry for measuring the strength and hardness of tablets. The operation of a standard friability tester was simulated using the discrete element method (EDEM™ from DEM Solutions). Simulations of a tablet coating pan at different pan speeds, pan loading and pan sizes were also performed using DEM. The predicted forces from the DEM simulations of both the systems were compared for various tablet shapes. It was observed that the average normal forces experienced by the tablets in the friability tester were 3 times larger than the average tangential forces. Moreover, the number of tablet/tablet collisions are fewer than the tablet/drum collisions for the standard number of tablets used in the tester (n=20). However, for large number of tablets (n=100), the number of tablet/tablet collisions are larger than the tablet/drum collisions. This situation is more representative of conditions in a large scale tablet coating pan. The forces in the coating pan generally showed the following trend: impact zone > shear layer > rotating layer. The force magnitudes were similar for all the tablet shapes that were investigated. The forces increased with pan speed, pan loading and pan size. The large scale pan showed ~ 1.5 to 4 times increase in the forces. Moreover, it was observed that the maximum normal force in the friability test was ~ 2 times greater than the maximum normal force in the large scale coating pan (pan diameter ~ 24"). However, the maximum tangential forces were similar. The results lead to a better understanding of the tester.

See more of this Session: Scale-up of Pharmaceutical Manufacturing Processes: Toward a QbD Approach
See more of this Group/Topical: Topical i : Comprehensive Quality by Design in Pharmaceutical Development and Manufacture

(108a) Implementation of Novel Integrated Pharmaceutical Processes: A Model-Based Approach

Alicia Román Martínez, Rafiqul Gani and John M. Woodley, Department of Chemical & Biochemical Engineering, Technical University of Denmark, Lyngby, Denmark

Although industrial biocatalysis has found a significant role in the manufacture of pharmaceuticals there are still significant obstacles to overcome. For example, unfavorable equilibrium and the product inhibition, as well as the difficult downstream processing, all affecting product yield. One way to overcome these issues is by integration of all reactions (chemical and enzymatic) in one-pot, and integration of the reaction with product recovery in one pot – in situ product removal (ISPR). Chemo-enzymatic processes involve combinations of at least two processing steps, like reactors and separators, where there are involved enzymatic bio-transformations coupled with another chemical or enzymatic reaction. Integration of this type of bio-chemical processes offers opportunities for improvements in terms of process performance, flexibility and better utilization of resources. However such processes have been hardly investigated comprehensively. Generally, design of this kind of processes, among other kinds, involves an iterative, trial and error experiment-based procedure where the experience of process designer plays an important role. Since experiments are usually time consuming and expensive, the search space of the potential designs needs to be significantly limited, lacking also a systematic approach to the process as a whole. Usually, a complete evaluation of all aspects is rarely achieved and sub-optimal processes may be introduced. This work focuses on presenting a model-based framework for design of these kinds of processes. This framework has the main following characteristics. First, all the potential options are considered via a superstructure to search the domain of potential process design; and second, it uses a decomposition approach to manage the complexity of the mathematical formulations representing complex optimization problems and therefore, the search space is reduced stepwise in order to locate candidate process options, giving an optimal design where further experimental efforts can be concentrated on. The application of this framework is described and highlighted with an important example in the pharmaceutical sector: the neuraminic acid synthesis. The model-based generic methodology was used to generate and evaluate the optimal process configuration which gives the optimum enzyme ratio and best purification method to maximize the product yield for the synthesis of this intermediate pharmaceutical, whose demand is increasing exponentially but the current methods of production are characterized by an unfavorable equilibrium, high amount of waste per kilogram and difficult downstream processing. The results show the relevance of one-pot reactions and ISPR techniques to achieve an improvement in the process performance.

See more of this Session: Model Based Integrated Design of Pharmaceutical Drug Product and Process
See more of this Group/Topical: Topical i : Comprehensive Quality by Design in Pharmaceutical Development and Manufacture

(108b) Design Space of Pharmaceutical Processes Using Data-Driven Based Methods

Fani Boukouvala, Fernando J. Muzzio and Marianthi G. Ierapetritou, Chemical and Biochemical Engineering, Rutgers University, Piscataway, NJ

The identification and graphical representation of process design space is very critical in locating not only feasible but also optimum operating variable ranges and design configurations. Design space of a process is the area of the parametric space within which acceptable product can be produced [1]. In the pharmaceutical industry- working towards an ideal maximum level of quality assurance and performance- the Design Space is directly linked to a well defined Control Strategy and Criticality Assessment [2]. In this work [3], it is proposed that the construction of the Design Space is considered as a problem of determining the boundaries of a process where feasible, profitable and acceptable performance of the process is guaranteed. For many years, this has been considered as a major concern in many process industries [4-7]. Specifically, the mapping of the design space of pharmaceutical processes is achieved using the ideas of process operability and flexibility under uncertainty. For this purpose, three different data-based methodologies are proposed to produce the graphical representation of the region in which acceptable performance of unit black- box processes can be ensured.

The first step of the proposed methodologies is the use of three different surrogate- based methodologies to model the performance of all possible designs of pharmaceutical unit operations. The methods used are: High Dimensional Model Representation (HDMR), Response Surface Methodology (RSM) and Kriging. In HDMR methodology the output of a process is considered as a finite hierarchical correlated function expansion, in terms of the independent and cooperative forms of input variables [8]. Look-up tables are then constructed so that component functions at any given arbitrary input variable can be interpolated. In RSM, for a problem containing n continuous input variables, an n-dimensional quadratic polynomial is used as the local model since quadratic behavior describes the mathematical geometry in the neighborhood of an optimum [9]. Based on Kriging methodology, the prediction is expressed as a weighted sum of the observed function values at sampling points that fall within a set interval around the point that is predicted [10]. A variance for each test point is also calculated, thus regions where subsequent sampling is required can be linked to a high variance.

Modeling and optimization of pharmaceutical processes often includes modeling using discrete decisions, which represent design variables such as the use, size or configuration of a specific part of a piece of equipment (e.g. screw size of a feeder, design of a nozzle, etc). In the published work, data-driven modeling of pharmaceutical processes often treats these types of variables as coded values based on which the final response surface is fitted. In the present work, however, individual models are produced for alternative process designs which are complemented by the assignment of a decision variable for each design. These approaches are compared and their efficiency is illustrated through two pharmaceutical case studies involving a continuous powder mixer, and a loss-in- weight feeder of a tablet production process.

The purpose of this work is to develop a general methodology that will construct a graphical representation of the region bounded by the limits within which acceptable product or process performance is achieved. For this, the only available knowledge on the system consists of a multivariate experimental data set of a desired output measured at different operating conditions for different design configurations. This is very often the problem faced in processes for which a physics- based model does not exist. The main advantages of the proposed methodology (independent of the specific data-driven approach used) are that (a) the accurate calculation of the corresponding range of input variables that result to acceptable performance, and (b) the design that results to optimum performance at different operating conditions can be identified. It should be also emphasized that the methodology proposed does not depend on the existence of first- principle models, and that the models developed are computationally inexpensive. Experimental validation of the results is also performed to verify the results of the proposed approaches. This work can be considered as a fundamental building block of a general methodology for defining the design space of black- box processes, which operate under uncertainty, and for which the only available information is an experimental data set.

References:

1. Lepore J and Spavins J, PQLI Design Space. Journal of Pharmaceutical Innovation, 2008; 3(2): 79-87.

2. Garcia T, Cook G, and Nosal R, PQLI Key Topics - Criticality, Design Space, and Control Strategy. Journal of Pharmaceutical Innovation, 2008; 3(2): 60-68.

3. Boukouvala F, Muzzio F, and Ierapetritou M. Design Space of Pharmaceutical Processes using Data-Driven based Methods. Submitted to Journal of Pharmaceutical Innovation, April 2010.

4. Banerjee I and Ierapetritou MG, Design Optimization under Parameter Uncertainty for General Black-Box Models. Industrial & Engineering Chemistry Research, 2002; 41(26): 6687-6697.

5. Floudas CA and Gumus ZH, Global Optimization in Design under Uncertainty: Feasibility Test and Flexibility Index Problems. Industrial & Engineering Chemistry Research, 2001; 40(20): 4267-4282.

6. Halemane KP and Grossmann IE. Optimal Process Design Under Uncertainty. 1987 [cited; Available from: http://hdl.handle.net/1903/4569

7. Lima F, Jia Z, Ierapetritou M, and Georgakis C, Similarities and differences between the concepts of operability and flexibility: The steady-state case. p. 702-716.

8. Li G, Rosenthal C, and Rabitz H, High Dimensional Model Representations. The Journal of Physical Chemistry A, 2001; 105(33): 7765-7777.

9. Myers RH and Montgomery DC, Response Surface Methodology: Process and Product in Optimization Using Designed Experiments: John Wiley \& Sons, Inc., 1995, 728.

10. Cressie N, Statistics for Spatial Data (Wiley Series in Probability and Statistics): Wiley-Interscience, 1993.

See more of this Session: Model Based Integrated Design of Pharmaceutical Drug Product and Process
See more of this Group/Topical: Topical i : Comprehensive Quality by Design in Pharmaceutical Development and Manufacture

(108c) Model Based Control of Drug Substance Particle Size to Ensure Drug Product Uniformity

Jon Hilden[1], Mark Schrad[1], Jessica T. Sloan[1], Jennifer Kuehne-Willmore[2] and Timothy Kramer[3], (1) Formulation and Process Development, Eli Lilly and Company, Indianapolis, IN, (2)Chemical Process Engineering, Eli Lilly and Company, Indianapolis, IN, (3)Statistics, Eli Lilly and Company, Indianapolis, IN

TBD - pending legal approval

See more of this Session: Model Based Integrated Design of Pharmaceutical Drug Product and Process
See more of this Group/Topical: Topical i : Comprehensive Quality by Design in Pharmaceutical Development and Manufacture

(108d) Modeling the Roll Compaction Process Using Finite Element Analysis

Ariel Muliadi[1], Carl R. Wassgren[1] and James D. Litster[2], (1)Mechanical Engineering, Purdue University, West Lafayette, IN, (2)Chemical Engineering, Purdue University, West Lafayette, IN

Roll compaction is part of a dry agglomeration process that takes place prior to tabletting. Its end product is agglomerate strips ('ribbons') that are ultimately milled for producing agglomerate granules. Owing to the granules' larger size, this process is more advantageous than direct compaction in many aspects, including reduction in dusting, improvement in flowability, and better die filling, all of which in turn makes capping less likely and production of high quality tablets more likely [1]. Recent experiments, however, have shown that density variations across the ribbon's width are common. Such variations in turn lead to a wide granule size distribution, which ultimately causes the process to lose its benefits.

The ribbon's non-uniform density distribution has been the subject of many previous studies. Those that took an experimental approach include Simon and Guigon [2, 3], who argued that the density distribution in the ribbon was the result of non-uniform material feed. Those that took a computational approach include Johanson [4] and Cunningham [5]. The former study was developed in the 1960s and was derived based on the empirical observation that roll pressure is a logarithmic function of ribbon density. Despite being attractive for its simplicity, this model provides only one-dimensional predictions and does not agree with many experimentally measured observations. The study by Cunningham [5] utilized the Finite Element Method (FEM) for two- and three-dimensional simulations of the roll compaction process. This study showed that FEM is capable of capturing the many characteristics of a roll compaction process. Quantitatively, however, the FEM predictions given in this study do not agree with experimental measurements. This discrepancy is likely because Cunningham did not account for the fact that mechanical properties of pharmaceutical powders are functions of their relative density.

This study revisits the FEM approach. However, unlike previous studies, the material mechanical properties—which are approximated using the Drucker-Prager Cap porous plasticity model—are varied depending on the local relative density using an external FORTRAN subroutine. To better approximate the roll compaction process, an Arbitrary Eulerian-Lagrangian (ALE) meshing algorithm is employed. This meshing strategy allows Eulerian boundaries to be defined at the feed (inlet) and exit (outlet) zones of the computational domain; essentially making the computational mesh act as a deformable control volume that allows material to enter from the inlet and exit from the outlet continuously. Further, an adaptive meshing scheme is also employed to accommodate for significant deformation.

The rollers and feed screw housing are modeled as un-deformable rigid bodies. Coulomb friction is assumed at the powder-roll interface. The effects of screw speed are approximated as a set pressure at the inlet zone. Methods used in approximating the inlet pressure will be discussed. Model outputs are validated experimentally by measuring the ribbon envelop density, as well as the normal and shear stresses acting on the ribbon. Normal and shear stresses predictions are also compared with Johanson's model. Reasons for discrepancies will be discussed.

References:

[1] Kleinebudde, P., 2004, "Roll compaction/dry granulation: pharmaceutical applications," European Journal of Pharmaceutics and Biopharmaceutics, 58, pp. 317-326.

[2] Simon, O., and Guigon, P., 2003, "Correlation between powder-packing properties and roll-press compact heterogeneity," Powder Technology, 130, pp. 257-264.

[3] Simon, O., and Guigon, P., 2003, "Roll press design--influence of force feed systems on compaction," Powder Technology, 130, pp. 41-48.

[4] Johanson, J. R., 1965, "A rolling theory for granular solids," ASME Journal of Applied Mechanics, E32(4), pp. 842-848.

[5] Cunningham, J. C., 2005, "Experimental studies and modeling of the roller compaction of pharmaceutical powders," PhD Thesis, Drexel University.

See more of this Session: Model Based Integrated Design of Pharmaceutical Drug Product and Process
See more of this Group/Topical: Topical i : Comprehensive Quality by Design in Pharmaceutical Development and Manufacture

(108e) Application of Model Based Drug Substance Particle Size Control in Drug Product Development

Mark Schrad[1], Jon Hilden[1], Jessica T. Sloan[1] and Jennifer Kuehne-Willmore[2], (1)Formulation and Process Development, Eli Lilly and Company, Indianapolis, IN, (2)Chemical Process Engineering, Eli Lilly and Company, Indianapolis, IN

TBD - pending legal approval

See more of this Session: Model Based Integrated Design of Pharmaceutical Drug Product and Process
See more of this Group/Topical: Topical i : Comprehensive Quality by Design in Pharmaceutical Development and Manufacture

(108f) Investigation of a Tablet Coating Process Using a Multi-Model Simulation Approach

Daniele Suzzi[1], Gregor Toschkoff[1], Daniel Machold[1] and Johannes Khinast[2], (1)Research Center Pharmaceutical Engineering (RCPE), Graz, Austria, (2)Institute for Process and Particle Engineering, Graz University of Technology, Graz, Austria

Spray coating is a frequently used process step in the pharmaceutical industry, benefits range from the control of the release of the active pharmaceutical ingredient (API) of a tablet to the masking of the taste of the API. One of the most common processes is drum coating, where the particles are placed in a rotating drum, and the coating liquid is sprayed onto the moving particle bed. A central quality attribute is the uniformity of the coating layer, as the coating usually has critical functional properties. Besides experimental work [1], numerical simulations have become an extremely important tool in process design.

In this work, we present an innovative approach to numerically analyze, understand and optimize important steps of the pharmaceutical tablet coating process. Coater performance depends on a range of very different parameters. Therefore, to find critical process parameters and determine their effect on quality, it is necessary to combine advanced numerical techniques on different scales. A central point is the application of the developed methods to industry-scale processes.

For example, for the simulation of the tablet bed mixing the Discrete Element Method (DEM) is used, while detailed numerical analysis of the interaction between spray droplets and tablet surface is done with advanced multiphase Computational Fluid Dynamics (CFD) methods [2]. On a larger scale, CFD can be adapted to investigate coater internal flow in terms of drying air and liquid spray movement. The results are used to optimize attributes like particles residence time, film formation or spray drying. Summarizing, the presented combination of simulation tools provide a deeper understanding of the different aspects of a coating process. The gained knowledge helps to go from trial-and-error based decisions to a Quality-by-Design approach.

[1] Dubey, A., Portillo, P.M., Muzzio, F.J., 2008. Study of Tablet Coating Thickness Uniformity Using LIBS, DEM and Compartment Modeling. 2008 AIChE Annual Meeting.

[2] Suzzi, D., Radl, S., Khinast, J.G., 2009. Local Analysis of the Tablet Coating Process: Impact of Operation Conditions on Film Quality. Submitted to Chem. Eng. Sci.

See more of this Session: Model Based Integrated Design of Pharmaceutical Drug Product and Process
See more of this Group/Topical: Topical i : Comprehensive Quality by Design in Pharmaceutical Development and Manufacture

(170a) Quality by Design for Generic Drugs: Overview

Robert A. Lionberger, Office of Generic Drugs, Food and Drug Administration, Rockville, MD

Quality by design is an essential part of the modern approach to pharmaceutical quality. But there is uncertainty in the generic drug industry about how the core concepts in ICH Q8(R1) will apply to development of generic drugs. FDA's Office of Generic Drugs has been revising its approaches to pharmaceutical quality in recent years, beginning with implementation of a question-based review process and continuing through current discussions on implementation of quality by design. In this talk, I will give examples about how the core elements of quality by design apply to generic drug development: starting with a Quality Target Product Profile, identifying critical quality attributes of the drug product, using prior knowledge to perform risk assessments for development prioritization, linking critical quality attributes to material attributes and process parameters and establishing a quality control strategy. There will be specific discussion on how some of the unique features of generic drug development make these processes different from how they may be employed in new drug development.

See more of this Session: Quality by Design in Generic Drug Development
See more of this Group/Topical: Topical i : Comprehensive Quality by Design in Pharmaceutical Development and Manufacture

(170b) Quality by Design in Pharmaceutical Development

Yatindra Joshi, Generics R&D, Teva Pharmaceuticals, North Wales, PA

More than 70% of the prescriptions in the US are now filled with generic products, and therefore, the industry has the responsibility to continue to provide safe, effective, and affordable medications. In 2009 GPhA initiated a dialogue between the Industry and the US FDA Office of Generics on Quality by Design (QbD) - a systematic approach for designing quality in our products. To facilitate the discussion between the two, FDA provided hypothetical cases on quality target product profile, compatibility between drug and excipients, formulation and process development, and scale-up and control strategies. While many questions still remain to be addressed, there is general agreement between the industry and the FDA on the benefit of quality by design. The purpose of this presentation is to provide the generic industry perspective on QbD.

See more of this Session: Quality by Design in Generic Drug Development
See more of this Group/Topical: Topical i : Comprehensive Quality by Design in Pharmaceutical Development and Manufacture

(170c) QbD for Generic Drugs: A Case Study for Immediate-Release Products

Zhigang Sun, Division of Chemistry II, Office of Generic Drugs, OPS, CDER, FDA, Rockville, MD

Quality by design (QbD) is a systematic approach for product and process design and development to ensure predefined product quality. Although the recent ICH guidelines (Q8-Q10) have provided an outline for development of "high quality" pharmaceutical products based on a comprehensive QbD approach, there are very few publications describing how to implement the QbD principles on the development of generic drug products.

This presentation will discuss a case study of the pharmaceutical development report for an immediate-release drug product, which was developed by the FDA/Industry Working Group in 2010. This example built on the Conformia ACE case study by treating that product as the reference product for development of a generic product. The purpose of this case study is to illustrate how to implement the principles contained in ICH guidelines (Q8 , Q9, and Q10) during the development of generic drug products.

See more of this Session: Quality by Design in Generic Drug Development
See more of this Group/Topical: Topical i : Comprehensive Quality by Design in Pharmaceutical Development and Manufacture

(170d) QbD for Generic Drugs: Modified Release Drug Products

Andre Raw, Office of Generic Drugs, Food and Drug Administration, Rockville, MD

The Office of Generic Drugs ensures the therapeutic equivalence and quality of generic drug product pharmaceuticals. This is based principally upon a regulatory paradigm that requires pharmaceutically equivalent drug products containing the same active ingredient, dosage form, and strength (among other things), to be shown to be bioequivalent to the brand name product. Also under this paradigm, the batch to batch quality of generic drug products is ensured based upon batch release testing to empirically derived specifications. While this approach has been used with great success toward bringing the availability of high quality of generic drug products, the vast majority of pharmaceuticals approved using this framework have been solution and immediate release oral products, which are inherently simple in design. Drug products however, have increased in design complexity, to encompass modified (oral) release products, transdermal delivery systems, among others complex dosage forms. As these pharmaceuticals have increased in their design complexity, approaches adopted to ensure the equivalence and quality of generic drug product pharmaceuticals, must likewise evolve.

FDA's Pharmaceutical cGMPs for the 21st Century: A Risk-Based Approach and Process Analytical Technology Initiatives are meant to enhance and modernize the regulation of pharmaceutical manufacturing and product quality. As a result of these initiatives and the trend of increasingly complex pharmaceuticals, OGD is currently implementing approaches to enhance the assessment of generic drug products, based upon a paradigm that places increased emphasis on "quality by design" as opposed to relying solely upon "quality by testing". Under this enhanced review system, in addition to relying upon "bioequivalence by testing" to the brand name, these increasingly complex generic drug products will be evaluated to ensure their "bioequivalence by design" based upon on an understanding of the critical quality attributes needed to provide an equivalent exposure profile essential toward ensuring equivalent clinical characteristics in the target patient population. This enhanced review framework also provides for a batch to batch quality control strategy, not solely upon batch release testing, but upon an understanding of formulation and process design, the application of process analytical technology for continuous monitoring, and where relevant, performance-based specifications.

See more of this Session: Quality by Design in Generic Drug Development
See more of this Group/Topical: Topical i : Comprehensive Quality by Design in Pharmaceutical Development and Manufacture

(170e) QbD in the Generic Pharmaceutical Industry: Anticipating Its Effects On the Development of Immediate and Modified Release Dosage Forms

John Kirsch, Product Development, Mylan Pharmaceuticals, Morgantown, WV

FDA and the generic pharmaceutical industry have been engaged in discussions regarding the implementation of specific quality by design principles since 2009. Though many elements of quality by design have long been utilized in generic product development, the expectations of FDA and an overall strategy for adoption of these principles in the generic industry remain unclear. This presentation will discuss the main elements of QbD which have been proposed for immediate release and modified release dosage forms, and will attempt to anticipate how these elements can be formally incorporated into generic product development.

See more of this Session: Quality by Design in Generic Drug Development
See more of this Group/Topical: Topical i : Comprehensive Quality by Design in Pharmaceutical Development and Manufacture

(170f) QbD in Generic Drug Development: PAT Application

Yanming Zu[1], Yi Luo[2] and **Salah U. Ahmed**[1], (1)Abon Pharmaceuticals, LLC, Northvale, NJ, (2)Teva Pharmaceuticals, USA, Pomona, NY

A generic drug product is pharmaceutically and therapeutically equivalent to a Reference Listed Drug (RLD) product. It meets the same requirements for identity, purity, quality, stability, and complies with the same rigid standards of GMP regulations for manufacturing as the innovator product. In addition, generic product development faces challenges of bioequivalence requirement and constraints of patent issues. In order to achieve the goals of QbD, FDA encourages the use of Process Analytical Technology (PAT) to promote innovation and efficiency in pharmaceutical development, manufacturing and quality assurance. The objective of PAT is to improve understanding and efficiently control the manufacturing process. The process understanding acquired by PAT will facilitate process design, control and optimization, justify and address the limitation of the time-defined end points and reduce the risk of failure during scale-up. Well-understood processes can be designed and developed during the product development stage to consistently ensure predefined product quality. PAT initiative provides valuable guidance that can be implemented during the development of generic products to achieve a robust manufacturing process. Information gained during process development will serve as a supporting tool for the implementation of QbD and address the questions in QbR. It also helps avoid manufacturing problems and aids in troubleshooting during the life span of a product. This presentation discusses the concept of PAT and its application in product development to support ANDA filing and approval. An overview of PAT application in typical manufacturing unit operations for oral solids and its benefits in applied pharmaceutical research are provided.

See more of this Session: Quality by Design in Generic Drug Development
See more of this Group/Topical: Topical i : Comprehensive Quality by Design in Pharmaceutical Development and Manufacture

QbD for generic drugs: Oral suspensions

Umesh R Pai, Sun Pharmaceutical Industries Inc., Cranbury, NJ, USA

Introduction

The paper focuses on the approach for systematic development of Oral suspensions. Suspension is a complex dosage form in which insoluble solid particles (internal phase) is dispersed uniformly throughout the suspending medium (external phase). Strategy for QbD in generic drug development of oral suspensions is presented.

Development Strategy

Develop Quality Target Product Profile

Upon identification of a oral suspension product for development, the formulation chemist will perform physical and chemical characterization of the reference listed drug product. A Quality Target Product Profile (QTPP) for the proposed product will be developed which represents the desired product characteristics of the generic product. Refer to Table 1 Quality Target Product Profile for Oral Suspension. The QTPP will be part of the product concept note that will be prepared by the formulation chemist which charters the development path. As knowledge is gained for the product during the development, the QTPP with the desired product characteristics will eventually evolve into the finished product and stability specifications for the final product.

Table 1. Quality Target Product Profile for Oral Suspension

Profile Component	Justification
Dosage form	Suspension, needs to match Reference Listed Drug
Strength/(s)	Needs to match Reference Listed Drug
Description / Appearance	Needed for acceptibility of the patient
Finished Product Performance	Product should meet predetermined finished product specification
Microbiology	Product should be stable from a microbiological perspective for the proposed shelf life
Packaging	Packaging should be non reactive and designed to contain the proposed deliverable volume as per the label
Bioequivalence	Should pass bioequivalence criteria as detailed in FDA guidance for C_{max}, $AUC_{0\text{-}t}$, $AUC_{0\text{-}inf}$
Stability Performance	Product should meet predetermined stability specification at accelerated and controlled room temperature stability condition, Related substances should meet ICH Q3B (R2) criteria

Derive Product Quality Attributes

The product quality attributes will be derived based on the QTPP. A preliminary list of potential product quality attributes derived from prior knowledge and/or the QTPP for oral suspension is presented in Table 2. Over the course of the development, those attributes that are critical (critical quality attributes, CQAs) to ensure the following target product profile requirements can be met will be identified. The list of product quality attributes will evolve in to controls when the formulation and manufacturing process are finalized and as product knowledge and process understanding increase.

Table 2. Identified Product Quality Attributes for oral suspension

List of Product Quality Attributes
Description / Appearance
Assay
Identification
Related compounds
pH
Uniformity of content
Weight/mL
Dissolution profile
Preservative content
Deliverable volume
Viscosity
Microbiology
Packaging
Bioequivalence

Perform Preliminary Risk Analysis

A preliminary risk analysis using cause and effect process prior to beginning of the development will be conducted. The risk assessment will be based upon the desired formulation characteristics and the manufacturing process. Prior knowledge, literature reviews and the desired quality product attributes will help propose mitigation strategies for the identified risks. These strategies will charter the developmental experiments. Failure Mode and Effect Analysis (FMEA) will be utilized for performing risk assessment and risk mitigation. The list of potential CQAs for subsequent evaluation during the development can be prioritized. FMEA will take in consideration the risks presented by a. raw material (Active pharmaceutical ingredient, Excipients) b. packaging material c. manufacturing process.

Perform Prototype Formulation Development

A thorough study of the drug substance which is the internal phase of the suspension and excipients will be performed with respect to their physicochemical properties. Preformulation studied will be performed to understand incompatibility of the drug substance with the excipients. The prototype formulation composition will be designed that can meet the required product quality attributes. Product container closure system will be designed. The critical material and functional attributes of the packaging container that can affect the product performance and provide protection to the drug product throughout its shelf life will be understood. Development stability studies will be conducted on the final prototyope formulation to understand performance of product under accelerated, intermediate and controlled room temperature stability conditions. Freeze-thaw cycling of the finished product will be conducted to understand the physical stability of the suspension. The product quality attributes that will be tested on stability are listed in Table 3. Stability indicating methods will be developed for testing of the related compounds. Product placed on stability will be evaluated for physical as well as chemical quality attributes. Rheoligical testing will be performed using cone and plate viscometer. Typically plots of Viscosity versus Shear Rate will be studied to understand rheological behaviour. Change in particle size, crystal habit and its impact on dissolution performance will be understood. The sedimentataion rate and sedimentation volume will be studied at time zero time and the stability conditions. The volume of sediment will be observed if it stays constant with time.

Table 3. Stability Product Quality Attributes for oral suspension

List of Product Quality Attributes
Description / Appearance
Assay
Related compounds
pH
Uniformity of content
Dissolution profile
Preservative content
Viscosity
Microbiology

Perform Formulation and Process Optimization Studies

Formulation and process optimization studies will be performed using Design of Experiments (DOE) and multivariate analysis. Multivariate experimental studies will be conducted utilizing DOE to establish relationship of material attributes and process parameters to product critical quality attributes. The results from these studies will be analyzed to determine the impact of process and raw material variability on the CQAs and thus, on the final product quality. Product hold time studies will be performed. The optimization and hold time studies will help design the manufacturing processes that can produce a quality product and ensure that it is robust. The critical process parameters (CPPs) and required process controls to monitor and ensure process is in control will be identified. The manufacturing process flow chart will be developed which represents the equipment, sequence of material addition, the process ranges and the in-process tests. Refer to Table 4 and 5 for list of formulation and process optimization variables respectively which will be considered for optimization studies.

Table 4. Formulation optimization variables for oral suspensions

Formulation optimization variables *	Studied output
Active drug substance particle size	Dissolution performance
Suspending agent	Sedimentation rate, Redispersibility
Wetting agent	Crystal growth
Buffering agent	pH and stability
Preservative	Product stable from a microbiological perspective

* Will be dependent on the formulation under development

Table 5. Process optimization variables for oral suspensions

Process optimization variables*	Studied output
Processing temperature	Product stability and viscosity
Mixing speed	Uniformity of content
Mixing time	Sedimentation rate, Redispersibility,

* Will be dependent on the formulation under development

Establish Control Strategy

An overall control strategy based on the product knowledge and process understanding will be developed. Specifications to control and ensure product quality will be developed from the results of the prototype development, development stability and optimization studies. The raw material specifications will be finalized based on the identified critical material attributes (CMA's). The predetermined finished product and stability specifications will be finalized which will be inclusive of the CQA's identified during the development and optimization experiments. The specifications will be set based on clinical relevance, science and risk based, and part of quality control strategy. Table 6 identifies the CQA's which will be inclusive of the finished product and stability specification.

Table 6. Identified Critical Product Quality Attributes for oral suspension

List of Critical Product Quality Attributes
Description / Appearance
Assay
Related compounds
pH
Uniformity of content
Dissolution profile
Preservative content
Viscosity
Microbiology

Perform Intermediate Risk Analysis

This constitutes the second stage of risk assessment and mitigation and will be conducted before the manufacture of the exhibit batch. FMEA will take in consideration the risks presented by the developed process at the bench and pilot scale and the potential risks prior to manufacture of an exhibit batch. The critical material attributes and the critical control parameters will be linked to the product critical quality attributes.

Exhibit batch manufacture and ANDA submission

The exhibit batch will be manufactured in manufacturing or pilot plant in equipment which is similar in design and principle with the proposed commercial equipment. The batch size will be at least about $1/10^{th}$ size of the proposed commercial scale batch. The exhibit batch finished product should meet the finished product acceptance criteria and demonstrate bioequivalence with the innovator. Stability studies will be conducted to confirm stability of the product at 3 month accelerated temperature and controlled room temperature stability conditions. The exhibit batch should meet predefined stability specifications. The ANDA will be submitted as per the FDA's regulations and requirements.

Commercialization

The final phase of risk assessment and risk mitigation will be conducted prior to the manufacture of the validation batches. FMEA will take in consideration the risks presented by the developed process at the pilot scale and the potential risks prior to manufacture of an process validation batch. If required, engineering batch of the size of the commercial batch can be manufactured for understanding the risks and confirming the proposed critical control parameters.

References

1. ICH Q3 B(R2) Impurities in New Drug Products, Final 8/4/ 2006

2. ICH Q8 (R2) Pharmaceutical Development, Final 11/20/ 2009

2. ICH Q9 Quality Risk Management, Final 6/1/ 2006

3. ICH Q10 Pharmaceutical Quality System, Final 4/5/ 2009

4. Q8, Q9, and Q10 Questions and Answers, Final 5/5/ 2010

(209a) QbD and the Transition From Wyeth to Pfizer for API Small-Molecule Development and Manufacture

John L. Dillon, API Technology Operations, Pfizer, Pearl River, NY

Quality by Design (QbD) is playing a major role in the development and manufacturing arenas of the Pharma industry. Process Understanding, the central component of QbD, is augmented by applying essential tools such as Multivariate Analysis (MVA), Design of Experiments (DOE) and Process Analytical Technology (PAT). While QbD in the Pharma industry is formally applied to the safety and efficacy of the product for the patient, the antecedent drug substance is a critical component of this process. A robust and well understood process for the production of drug substance evolves from a risk-based approach for defining critical quality attributes and establishing the design space for the API, key intermediates and starting materials to assure consistent quality. Process understanding begins during development, so it is imperative that an effective system is in place to impart QbD that is effectively bridged to manufacturing and product lifecycle within an environment of co-development. In addition to a sound, science-based approach, QbD is an important component of good business practice leading to fewer surprises during manufacturing, thereby affording good supply assurance. The application of QbD in API small molecule development and manufacturing during the transition from legacy Wyeth to Pfizer will be presented within the framework of the different models used at each company. Brief case studies will be utilized to demonstrate application of QbD for API's for a late development drug candidate and a marketed product.

See more of this Session: Challenges Facing CMC Organizations in the Changing Business and Regulatory Environment
See more of this Group/Topical: Topical i : Comprehensive Quality by Design in Pharmaceutical Development and Manufacture

(209b) Advancing the Abbott Pipeline

Howard E. Morton, GPRD Process R&D, Abbott Laboratories, North Chicago, IL

The Pharmaceutical industry has been facing significant challenges due to rising R&D costs, increased regulatory expectations and fewer new product approvals. In this environment, increasing probability of success and process knowledge while at the same time reducing cycle time and containing cost calls for new strategies across the entire API development timeline. In early Development our business model encompasses interfacing closely with our partners in Development (Preclinical Safety, DMPK) and Discovery to enhance both compound selection and interrogation to allow timelier and more informed decision making. As an API proceeds through development essential process understanding is achieved utilizing a Quality by Design (QbD) approach employing Process Analytical Technologies (PAT), Modeling and Design of Experiments (DOE) protocols. A key focus has been the adoption of a "Material Science" strategy with the Drug Product group for the co-development of the final API form including the understanding and manipulation of the API physical property attributes that impact the final drug product. Brief examples will be given highlighting our QbD and Material Science efforts.

See more of this Session: Challenges Facing CMC Organizations in the Changing Business and Regulatory Environment
See more of this Group/Topical: Topical i : Comprehensive Quality by Design in Pharmaceutical Development and Manufacture

(209c) Commercialization and Quality by Design: Towards An Improved Model for Pharmaceutical Development, Launch and Supply

Michael P. Thien, Global Science, Technology & Commercialization, Merck & Co., Inc., Rahway, NJ

Five years ago, in response to unprecedented and multifocal challenges in the pharmaceutical industry, Merck adopted a new business model and approach for the development and supply of our products. This presentation will share how we have moved forward with a new Global Pharmaceutical Commercialization model integrating Lean Six Sigma and Quality by Design (QbD) to meet these challenges and enable us to develop new medicines providing value to patients around the world.

In the past we often relied on a single point of transfer—a knowledge handoff—encompassing key activities between divisions, on the critical path to pharmaceutical product registration. The commercialization model sought to provide industry leading cycle times, productivity enhancements on a per program basis, and optimization of the variable/fixed cost ratio by more strategically leveraging external capabilities and capacity. Through the commercialization model, we have been able to recognize these benefits by bringing manufacturing closer to development, optimizing timing of capital investments to avoid significant expenditures where product potential is uncertain, executing a product portfolio strategy that focuses on global optimum solutions that maximize business and patient value, and transforming our culture to meet these goals.

Schematic of Global Pharmaceutical Commercialization model . Commercialization manufacturing sites are part of the development and launch phases for a new product.

THE GRAPHIC ON THIS PAGE WAS NOT AVAILABLE ON THE E-MEDIA VERSION OF THIS PAPER.

The question we had to ask and answer was "What is the best way to execute on the Commercialization model? How can we pull together all of the Commercialization elements in a customer-focused way and with a common framework to truly change how we work on a day to day basis?". The answer was Quality by Design (QbD), but a QbD approach augmented with Lean-Six Sigma methodology. QbD, in our view, naturally required an integrated approach to product and process development, fully aligned with our Commercialization model's aspirations. It also could provide a common framework and nomenclature from development to supply and do so in a systematic and focused way. The Merck QbD roadmap was consequently constructed with the goal of realizing the opportunities offered by QbD throughout the product's lifecycle. In development, we looked to the reduction of development costs through the use of prior knowledge, technology platforms including PAT, and focus on cross-functional optimization to maximize value of process development while minimizing overall work. Once a marketed product, we sought goals of reducing atypicals, continuous improvement of the process and overall enhancement of customer focus.

To date, we have greatly exceeded the target of deferring capital until after commercial launch. We have demonstrated a significant productivity increase for technical resources in our global pharmaceutical organization vs. our pre-2005 baseline while simultaneously supporting a pipeline that is appreciably larger. Quality by Design drove the use of the enhanced risk management/risk based development approaches that have resulted in reduced costs while improving process capability (for API, Pharm and Analytical) as measured by CPK. Today, all new products are developed within the Commercialization paradigm, following the QbD paradigm.

Commercialization and QbD (with its Lean Six Sigma basis) continue to be foundational to our future success. It has provided a clear business process for getting global optima solutions codified in the development arena, and laid the foundations for feed forward and feed back processes across the development space and into supply. The commercialization model has allowed us to create an integrated virtual organization and a global technical community. Commercialization has also driven a culture where continuous improvement, of processes, and systems positions the organization to meet ongoing and future challenges in the industry.

See more of this Session: Challenges Facing CMC Organizations in the Changing Business and Regulatory Environment
See more of this Group/Topical: Topical i : Comprehensive Quality by Design in Pharmaceutical Development and Manufacture

(209d) Creating a Capabilities Driven Organization: Efficient Design of API Processes

Bret E. Huff, Chemical Product R&D, Eli Lilly and Company, Indianapolis, IN
See more of this Session: Challenges Facing CMC Organizations in the Changing Business and Regulatory Environment
See more of this Group/Topical: Topical i : Comprehensive Quality by Design in Pharmaceutical Development and Manufacture

(299a) Question-Based Review for Generic Drugs

Lawrence Yu, Office of Generic Drugs, FDA CDER, Rockville, MD

The Food and Drug Administration Office of Generic Drugs (OGD) has developed and implemented Question-based Review (QbR) with the broad goal of encouraging generic applicants to include quality-by design (QbD) principles in their product and process development. In light of this goal, OGD outlined the following four major benefits: • Ensure product quality through product design and performance-based specifications • Facilitate continuous improvement and reduce CMC supplements through risk assessment • Enhance the quality, transparency, and consistency of reviews through standardized review questions • Reduce CMC review time when applicants submit a quality overall summary that adequately addresses the QbR questions.

This presentation will provide an assessment of how QbR has contributed to achieving these benefits. It will identify the current approach's strengths and areas for improvement.

See more of this Session: Quality-by-Design (QbD) Round Table Discussion Session: An Integrated and Science-Based Approach for Quality by Design (QbD) Implementation in Pharmaceutical Industry
See more of this Group/Topical: Topical i : Comprehensive Quality by Design in Pharmaceutical Development and Manufacture

(299b) QbD at the Boundary

San Kiang, Process R&D, Bristol-Myers Squibb Company, New Brunswick, NJ

Quality Risk Assessments (QRA) for API and DP processes are mostly performed independently to examine the chemical or formulation processes themselves in terms of process risk and potential process control strategies. However, many of the process risks in a formulation process can be attributed to the properties of the API. In my talk, I will examine the potential application of QRA's to examine the overall risk across the API-DP boundary.

See more of this Session: Quality-by-Design (QbD) Round Table Discussion Session: An Integrated and Science-Based Approach for Quality by Design (QbD) Implementation in Pharmaceutical Industry
See more of this Group/Topical: Topical i : Comprehensive Quality by Design in Pharmaceutical Development and Manufacture

(299c) Implementing Quality by Design (QbD) for New Drug Products - A Regulatory Perspective

Sharmista Chatterjee, Ondqa/Cder/FDA, FDA, Silver Spring, MD

Quality by Design (QbD) is a science based approach for pharmaceutical development and manufacturing that begins with defining clinically relevant product attributes and then designing and implementing a process to consistently deliver desired quality product. QbD continues to be a major goal in the agency's initiative to encourage modernization of pharmaceutical manufacturing. Since the launch of the CMC Pilot Program by the Office of New Drug Quality Assessment (ONDQA) in July 2005, the office has had an opportunity to evaluate many QbD based submissions submitted both within or outside of the CMC Pilot Program. The objective of this presentation is to provide a regulatory insight about the current state of implementation of QbD, highlighting some of the benefits as well as the challenges. In addition, a regulatory perspective will be offered regarding considerations for successful implementation of QbD throughout the product life cycle.

See more of this Session: Quality-by-Design (QbD) Round Table Discussion Session: An Integrated and Science-Based Approach for Quality by Design (QbD) Implementation in Pharmaceutical Industry
See more of this Group/Topical: Topical i : Comprehensive Quality by Design in Pharmaceutical Development and Manufacture

(299d) The Heartburn Associated with Implementing Quality by Design Principles

Kevin D. Seibert, Chemical Product Research & Development, Eli Lilly & Co., Indianapolis, IN

Many challenges exist with implementation of the various aspects of Quality by Design. Challenges include the integration of newer approaches to development and technology transfer into facilities running long-established processes developed and installed using more traditional approaches. Additionally, Quality Systems and procedures that have long been in place require modernization to allow for flexible processing approaches and the use of advanced control mechanisms utilizing PAT. Lastly, understanding the potential changes to regulatory commitments and the resulting changes to notification requirements, and the potential cost/benefits to the pharmaceutical industry is not yet fully understood and tested.

While differences in approaches are clearly apparent between various pharmaceutical companies, a more difficult challenge exists when differing opinions and interpretations exist within the same company across different divisions. Small molecule interpretations and drivers may be significantly different than those in the large molecule areas. Differences in drug product and drug substance manufacturing, and the inherent differences in the unit operations, starting material variability, control systems, ability to build representative models, and the ability to perform designed experiments at representative scales also poses some significant barriers and challenges to the implementation of Quality by Design Principles.

See more of this Session: Quality-by-Design (QbD) Round Table Discussion Session: An Integrated and Science-Based Approach for Quality by Design (QbD) Implementation in Pharmaceutical Industry
See more of this Group/Topical: Topical i : Comprehensive Quality by Design in Pharmaceutical Development and Manufacture

(299e) Quality by Design and Mechanistic Models of Pharmaceutical Unit Operations

Pavol Rajniak[1], Shane Grosser[2], John Lepore[2], Rey Chern[1] and Eric Ahuja[3], (1)Pharmaceutical Commercialization Technology, Merck &Co., West Point, PA, (2)Chemical Process Development & Commercialization, Merck &Co., Rahway, NJ, (3)Analytical Chemistry Development & Supply, Merck &Co., West Point, PA

Developing pharmaceutical product formulations in a timely manner and ensuring quality is a complex process. Within the spirit of the QbD initiative it requires a systematic, science based approach and collection and combination of information from various categories, including properties of drug substance and excipients, interactions between materials, unit operations, and equipment. Advances in measurement tools in combination with quantitative analysis techniques (e.g. statistical tools, empirical correlations or mechanistic models) enable engineers to design and scale-up pharmaceutical unit operations and processes with more confidence and reliability. Manufacture of most drugs employs different (usually batch) unit operations of the drug substance manufacture or the drug product manufacture which can be categorized as follows: A/ Unit operations of the drug substance manufacture (reactions, separations, crystallization of the active pharmaceutical ingredient, API) B/ Unit operations of the sterile liquid products manufacture (mixing, dissolution of powders, filtration, lyophilization) C/ Unit operations of the solid dosage products manufacture (milling, blending, conveying, granulation, tabletting, encapsulation) Mathematical modelling of unit operations of the group A is common in chemical engineering community and commercial software exists for batch reactors, separators or crystallizers. Mathematical modelling of unit operations of group B is less common. Commercial software is not available but general level of modeling and understanding is already at satisfactory level. Mathematical modeling and understanding of unit operations of group C is less rigorous and more challenging than for groups A and B. A key reason is still unsatisfactory understanding of multi-phase gas-solid flow, particularly for highly concentrated small, non-spherical or cohesive particles, crystals, powders or granules. Depending on the model complexity, different approaches of the model implementation and dissemination must be used. The goal is to develop user-friendly models, preferentially using the environment common for collection of experimental data and allowing equation-solving and visualization of simulation results (Excel, Dynochem, Aspen, Visual Fortran, Fluent). Goal of the presentation is to share our experience how to use models for development and scale-up and to initiate discussion about the current practice and strategy of applications of mechanistic models within the QbD process: 1/ if, when and how to employ models in the QbD filing for the product registration? 2/ what level of the model verification and validation is needed for its accreditation by FDA? 3/ how to extend the model use from development to manufacture and supply?

See more of this Session: Quality-by-Design (QbD) Round Table Discussion Session: An Integrated and Science-Based Approach for Quality by Design (QbD) Implementation in Pharmaceutical Industry

(299f) Quality-by-Design (QbD): Integration of Real-Time PAT Process Monitoring and off-Line Product Characterization to Explore the Linkage Between a Pharmaceutical Process and Product

Huiquan Wu, Division of Product Quality Research, HFD-940, FDA, Silver Spring, MD and Mansoor A. Khan, Cder/Ops/Otr/Dpqr, Food and Drug Administratin, Silver Spring, MD

Real-time PAT process monitoring offers an unprecedented opportunity to observe what is going on inside the process and thus provides a window to study the process dynamics. Off-line characterization of the product could provide information about the physical/chemical properties of the product. Although from a process engineering perspective, the effects of process conditions on the product quality exist anyway for any process, the linkage between the process conditions and product quality attributes may not be obvious or easily appreciable. The lacking of easily appreciable linkage may present a barrier for the implementation of PAT and QbD in the pharmaceutical industry. In this work, several case studies will be presented to explore this linkage and its regulatory utilities in the pharmaceutical CMC area.

See more of this Session: Quality-by-Design (QbD) Round Table Discussion Session: An Integrated and Science-Based Approach for Quality by Design (QbD) Implementation in Pharmaceutical Industry
See more of this Group/Topical: Topical i : Comprehensive Quality by Design in Pharmaceutical Development and Manufacture

(299g) Quality-by-Design (QbD) Round Table Discussion

Christine Seymour, Pfizer Global Research & Development, Pfizer, Groton, CT and Huiquan Wu, Division of Product Quality Research, HFD-940, FDA, Silver Spring, MD

Round Table Discussion

See more of this Session: Quality-by-Design (QbD) Round Table Discussion Session: An Integrated and Science-Based Approach for Quality by Design (QbD) Implementation in Pharmaceutical Industry
See more of this Group/Topical: Topical i : Comprehensive Quality by Design in Pharmaceutical Development and Manufacture

Design Space and QbD:

It's All about the (Stochastic) Distributions

John J. Peterson
GlaxoSmithKline Pharmaceuticals
john.peterson@gsk.com

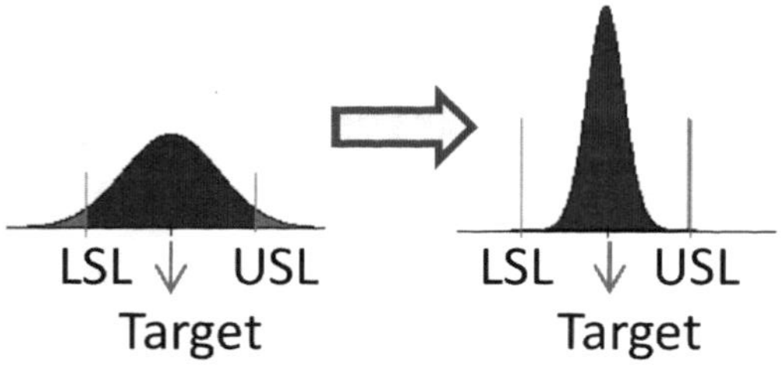

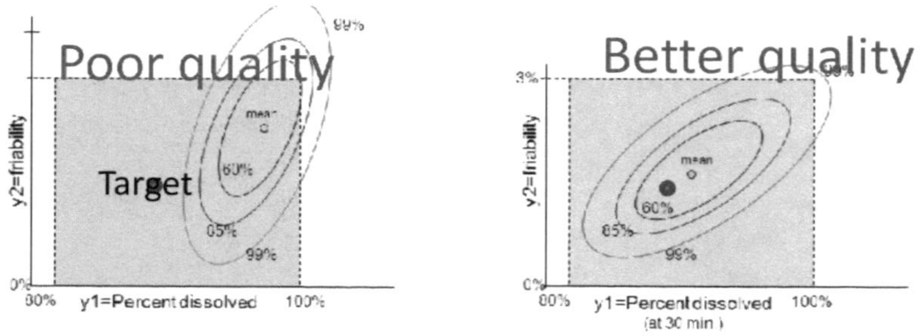

AIChE Annual Meeting – Salt Lake City, UT, November 9th, 2010

What is *quality improvement*?

- It's been said that Quality Improvement is about reduction in ***variation*** about a target (e.g. Montgomery, 2009, *Introduction to Statistical Quality Control*)

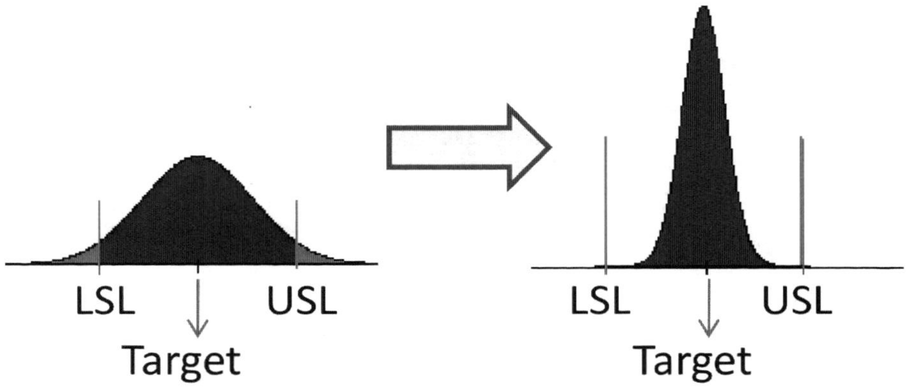

Quality Improvement – the strategic reduction of variation about a target

- Distribution for one quality response:

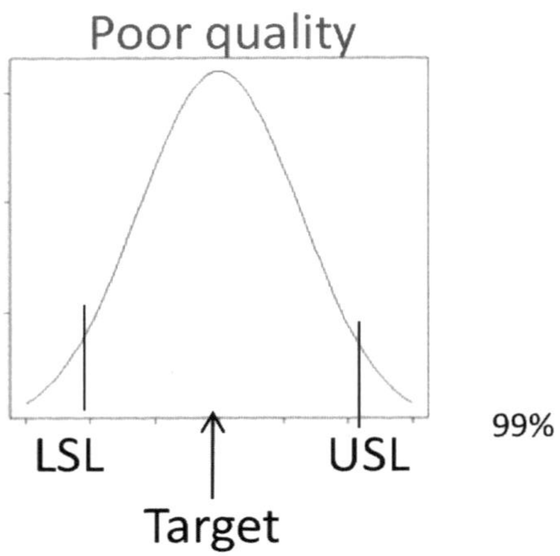

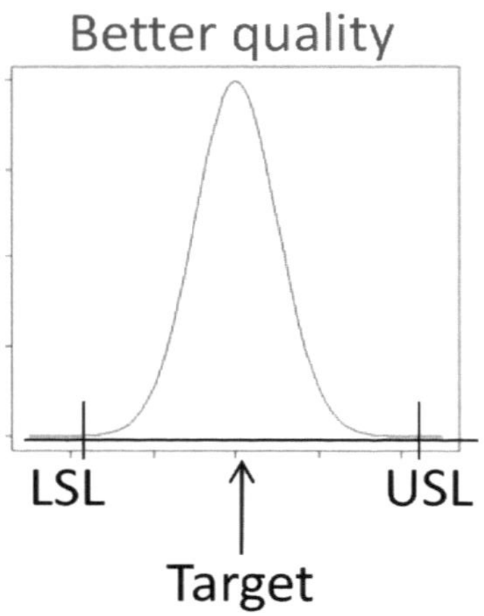

Reduction in *random* variation about a target

- Distribution for two quality responses:

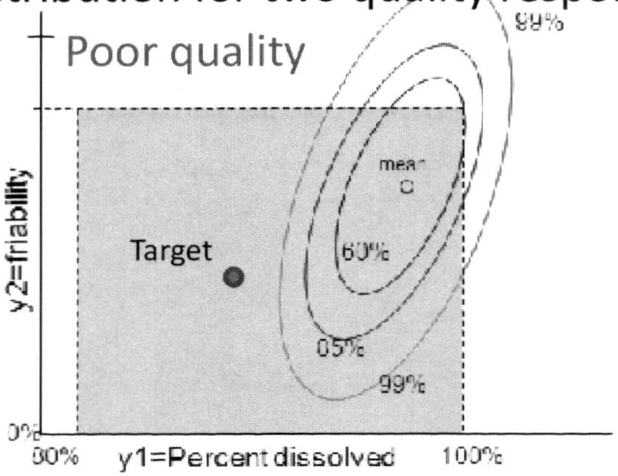

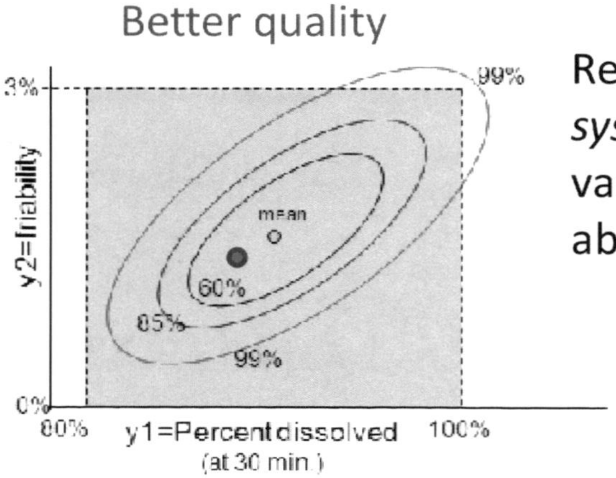

Reduction in *systematic* variation about a target

QbD and ICH Q8 Design Space

The ICH Q8 FDA Guidance for Industry defines "Design Space" as:
"The multidimensional combination and interaction of input variables (e.g. material attributes) and process parameters that have been demonstrated to provide assurance of quality."

- *Three key concepts:*

1. **Measurement**
   For example: controllable factors, input material attributes, in-process measurements, quality response measurements.

2. **Prediction**
   - Models to relate the predictive measurements to the quality responses. These need to be compared to specifications for quality.
   - Mean predictions are not enough! Predictive ***distributions*** are necessary.

3. **Reliability (to quantify risk)**
   To quantify "How much assurance?"

- The QbD-oriented guidance (PAT, ICH Q8, Q9, Q10, etc) is inundated with the words "risk" and "risk-based".)
- See presentation by H. Gregg Claycamp (CDER), "Room for Probability in ICH Q9"

Measurements

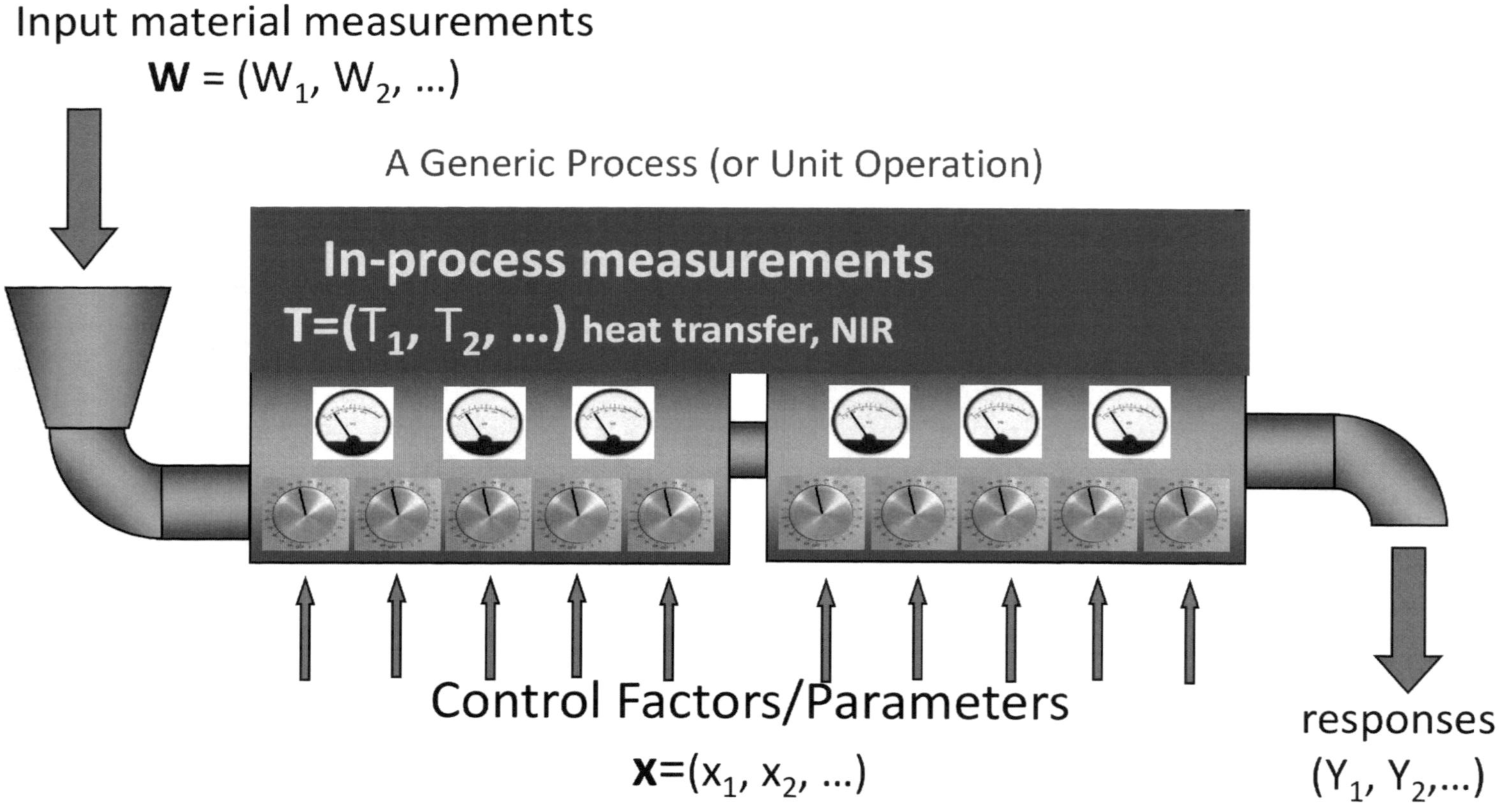

- Responses have specification limits which define Quality: $\{ A_{iL} < Y_i < A_{iU} \}$
- Vector of predictive variables, $(\mathbf{x}, \mathbf{z}) = (\mathbf{x},\mathbf{w},\mathbf{t})$ ($\mathbf{z}=(\mathbf{w},\mathbf{t})$, $\mathbf{z}$'s are noisy)

Prediction and Reliability

1. Empirical or mechanistic models typically are used to make predictions about future response values for a specified set of process conditions.

2. Any complex process that is shut down and restarted from scratch (using different batches of starting materials, etc.) to produce a true replicate response value, will produce **different** responses even under "identical" operating conditions (even if using extremely accurate measuring devices).

3. So point 2 (above) implies that we do **not** have the following simplistic model concept:

"predicted response"= "true model function" + "measurement error".

Prediction and Reliability

2. Any complex process that is shut down and restarted from scratch (using different batches of starting materials, etc.) to produce a true replicate response value, will produce **different** responses even under "identical" operating conditions (even if using extremely accurate measuring devices).

3. So point 2 (above) implies that we do **not** have the following simplistic model concept:

"predicted response"= "true (deterministic) model function" + "measurement error".

4. Instead, we have the follow model concept:

"distribution of predicted responses" = "stochastic process" + "measurement error".

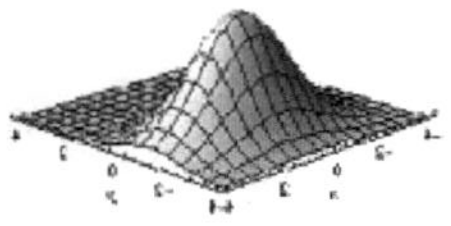

Focusing only on the mean (average) can put us at risk!

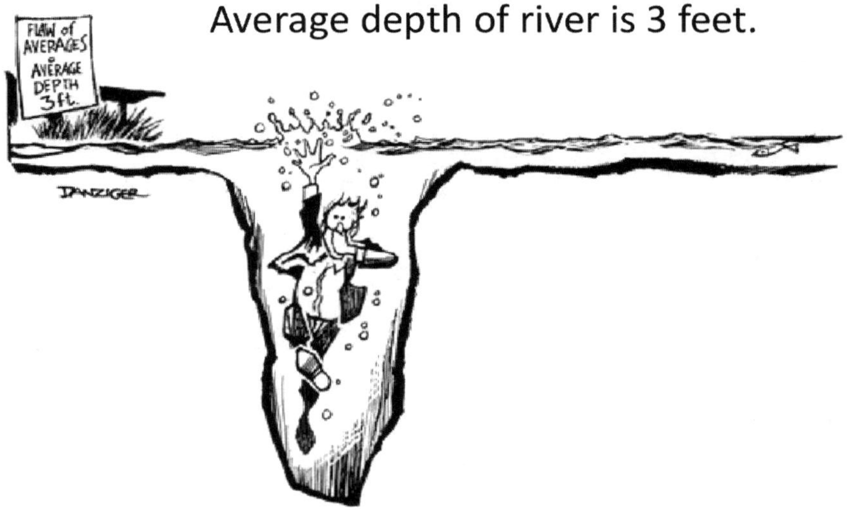

Average depth of river is 3 feet.

The Flaw of Averages:
Why We Underestimate Risk in the Face of Uncertainty
by Dr. Sam Savage

Prediction and Reliability

- **Predictive** *Distributions* are an important concept as they can be used to quantify the **Reliability** of meeting quality specifications.

- This is important "...to provide assurance of quality" as required by the ICH Q8 definition of design space.

  The ICH Q8 FDA Guidance for Industry defines "Design Space" as:

  "The multidimensional combination and interaction of input variables (e.g. material attributes) and process parameters that have been demonstrated to provide assurance of quality."

- ICH Q8 definition also begs the question: **"How much assurance?"**

Prediction and Reliability

- ICH Q8 definition also begs the question: **"How much assurance?"**

- This question can be answered in a quantitative, scientific fashion using a reliability (i.e. probability) measure.

- The QbD-oriented guidance (PAT, ICH Q8, Q9, Q10, etc) is inundated with the words "risk" and "risk-based".)

- But to quantify "risk" you need a probability measure.

- We compute probability measures from probability *distributions*.

- See also the presentation by H. Gregg Claycamp (FDA-CDER), "Room for Probability in ICH Q9"

Predictive Distributions and Multiple Response Process Optimization

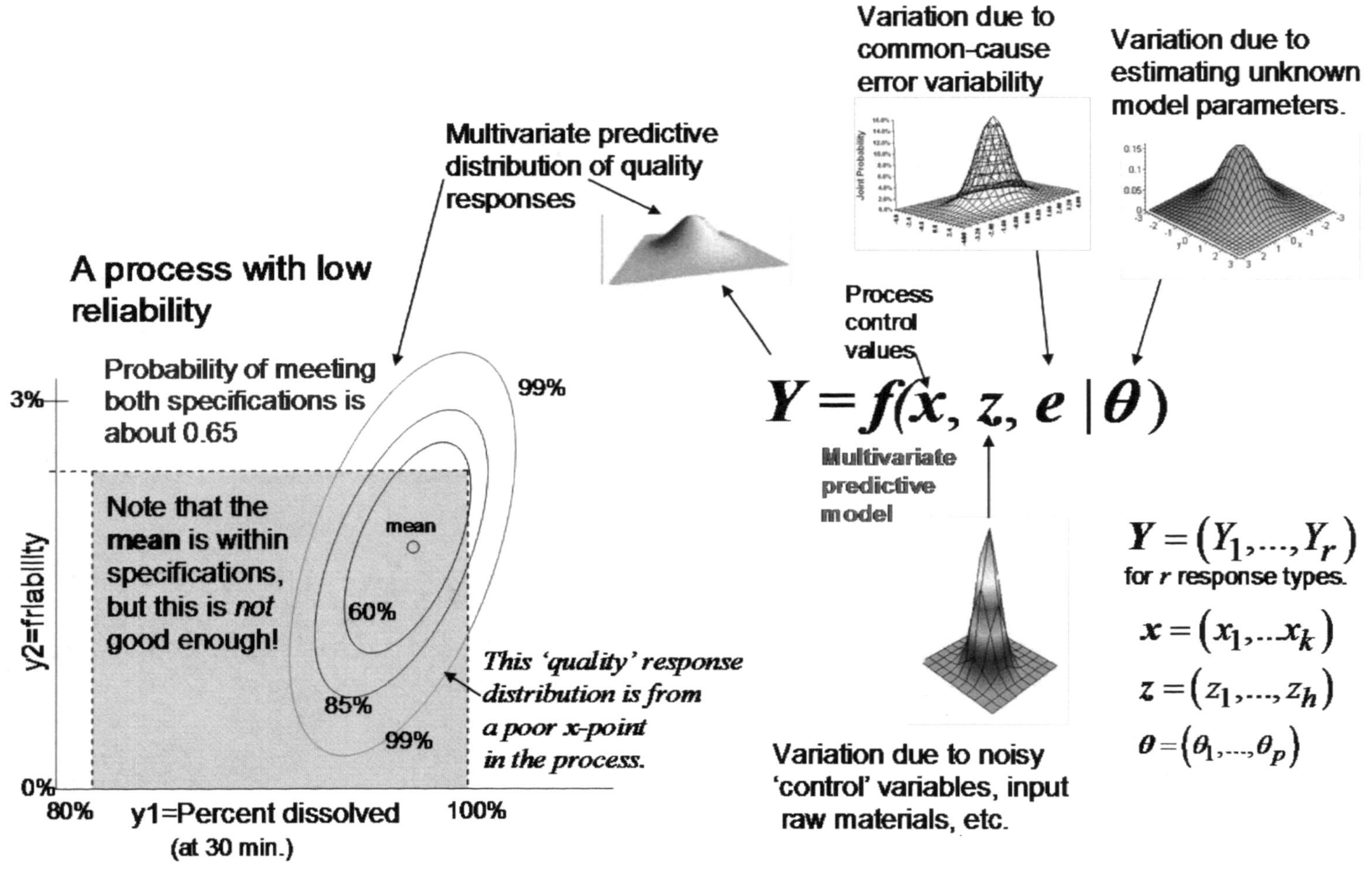

Predictive Distributions and Multiple Response Process Optimization

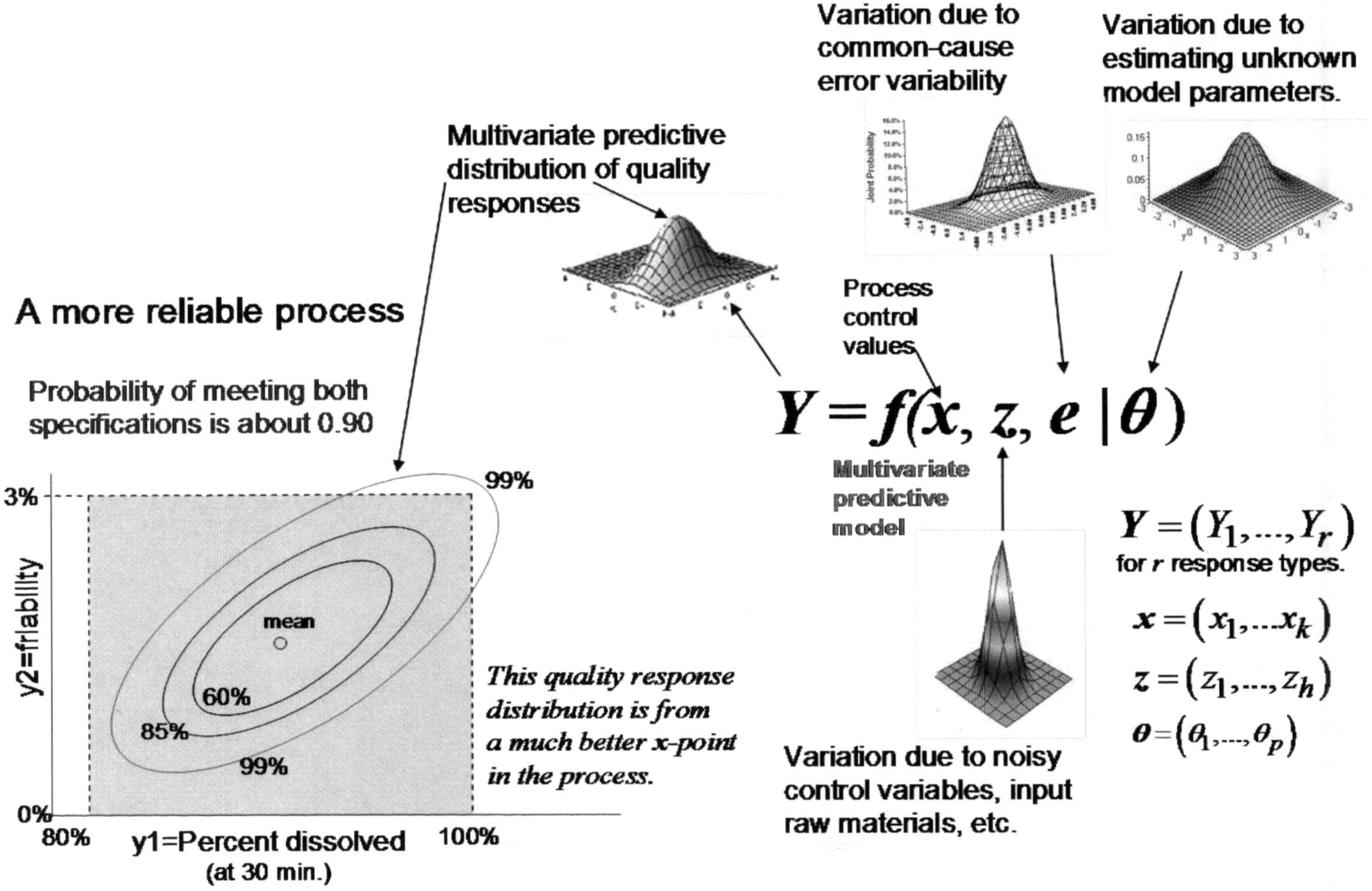

Predictive Distributions and Multiple Response Process Optimization

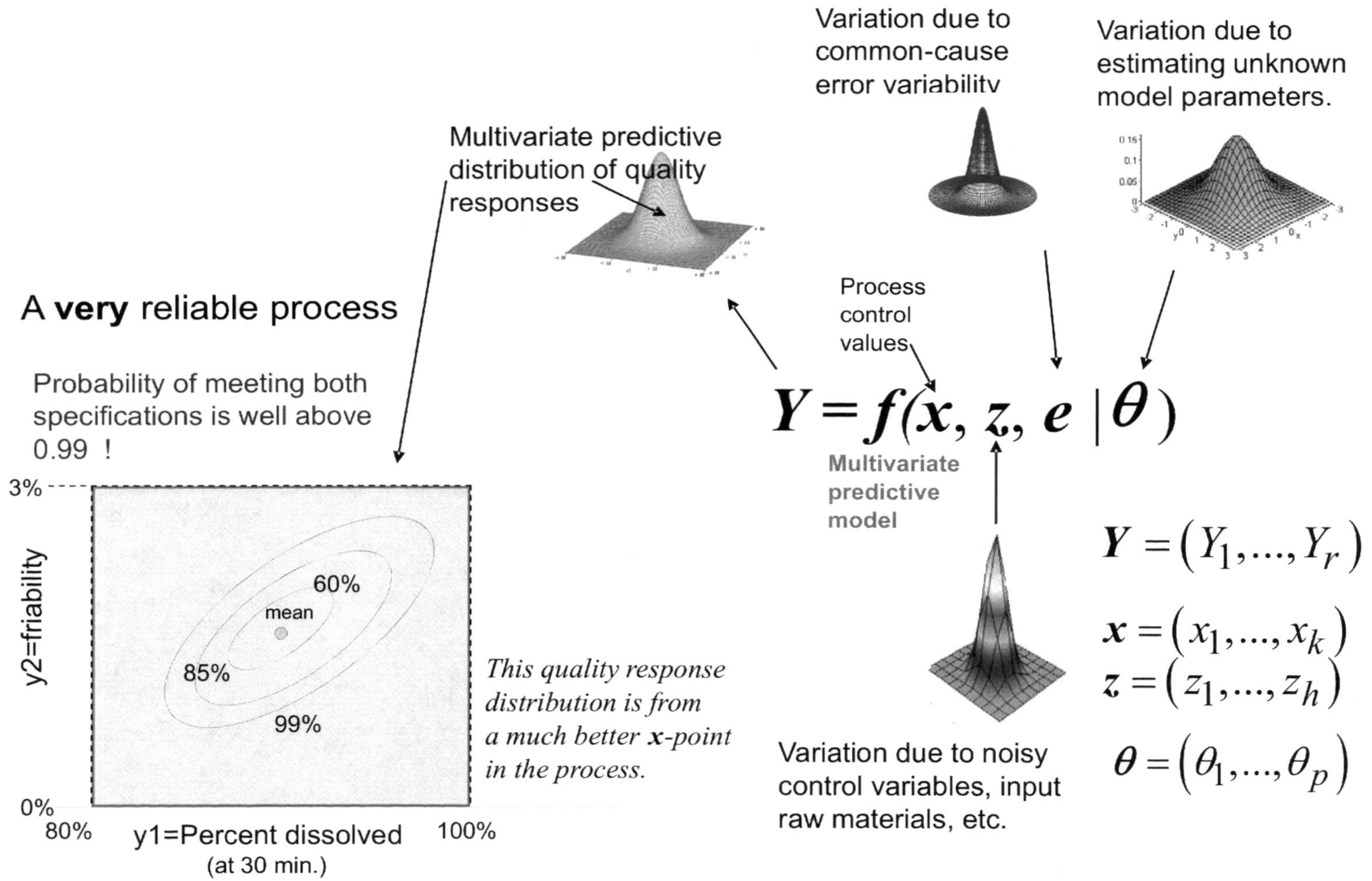

So…how can we improve quality and construct a reliable design space?

1. Build a *stochastic* model of your process.
   a. This can be a mechanistic or empirical model, as long as it predicts well.
   b. This stochastic model should produce a ***distribution*** of predicted values.

2. Through sequential (and designed) experimentation, find conditions to reduce process variation about the target.

3. The design space is the set of all process conditions (and inputs) that are associated with acceptably small variation (i.e. high reliability) about the target.

PS

Don't forget to model the uncertainty of your *unknown* model parameters!

How do we construct a predictive distribution for our process?

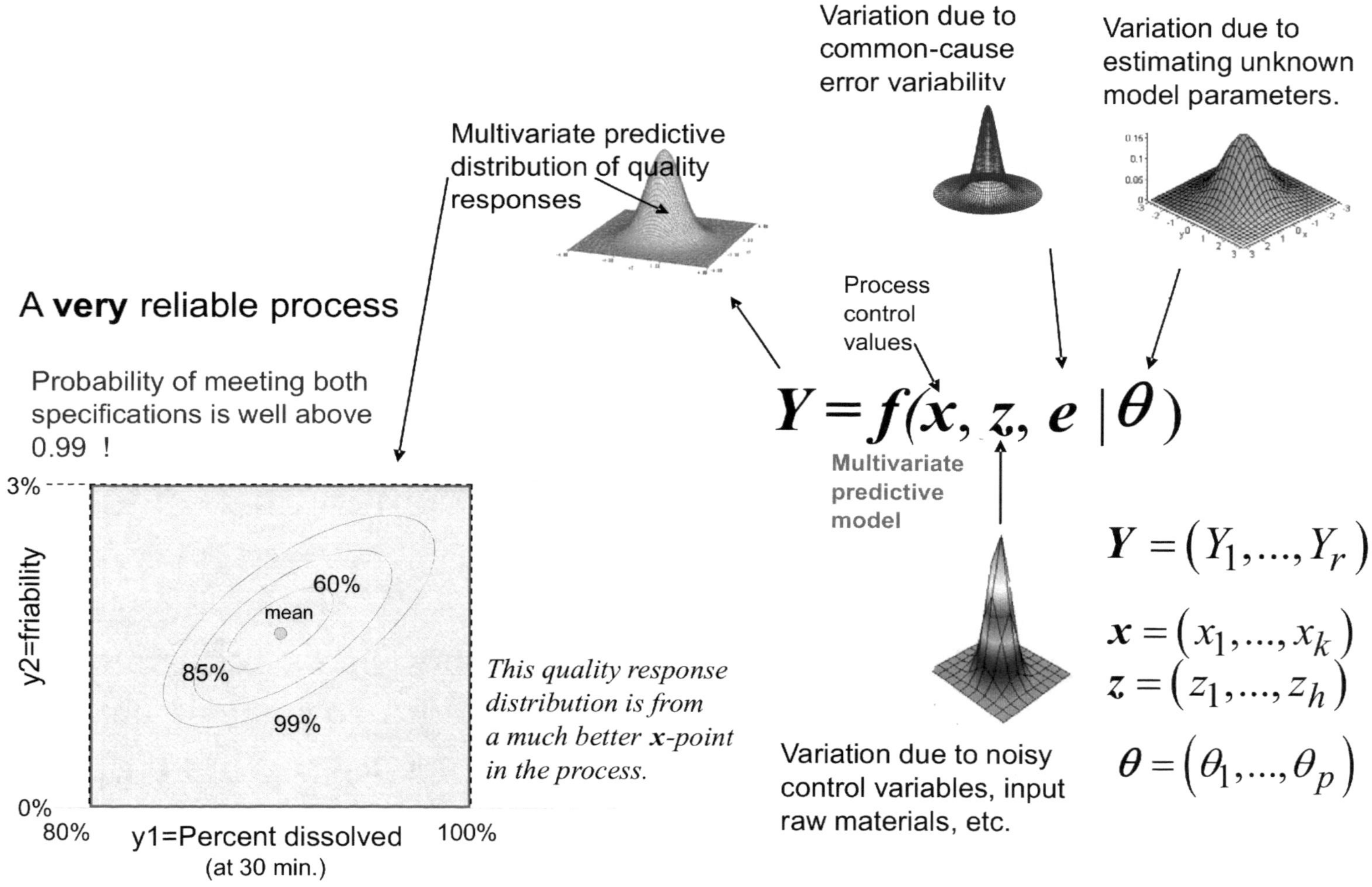

$$Y = f(x, z, e \mid \theta)$$

$$Y = (Y_1, \ldots, Y_r)$$

$$x = (x_1, \ldots, x_k)$$

$$z = (z_1, \ldots, z_h)$$

$$\theta = (\theta_1, \ldots, \theta_p)$$

How do we construct a predictive distribution for our process?

Use Monte Carlo simulations

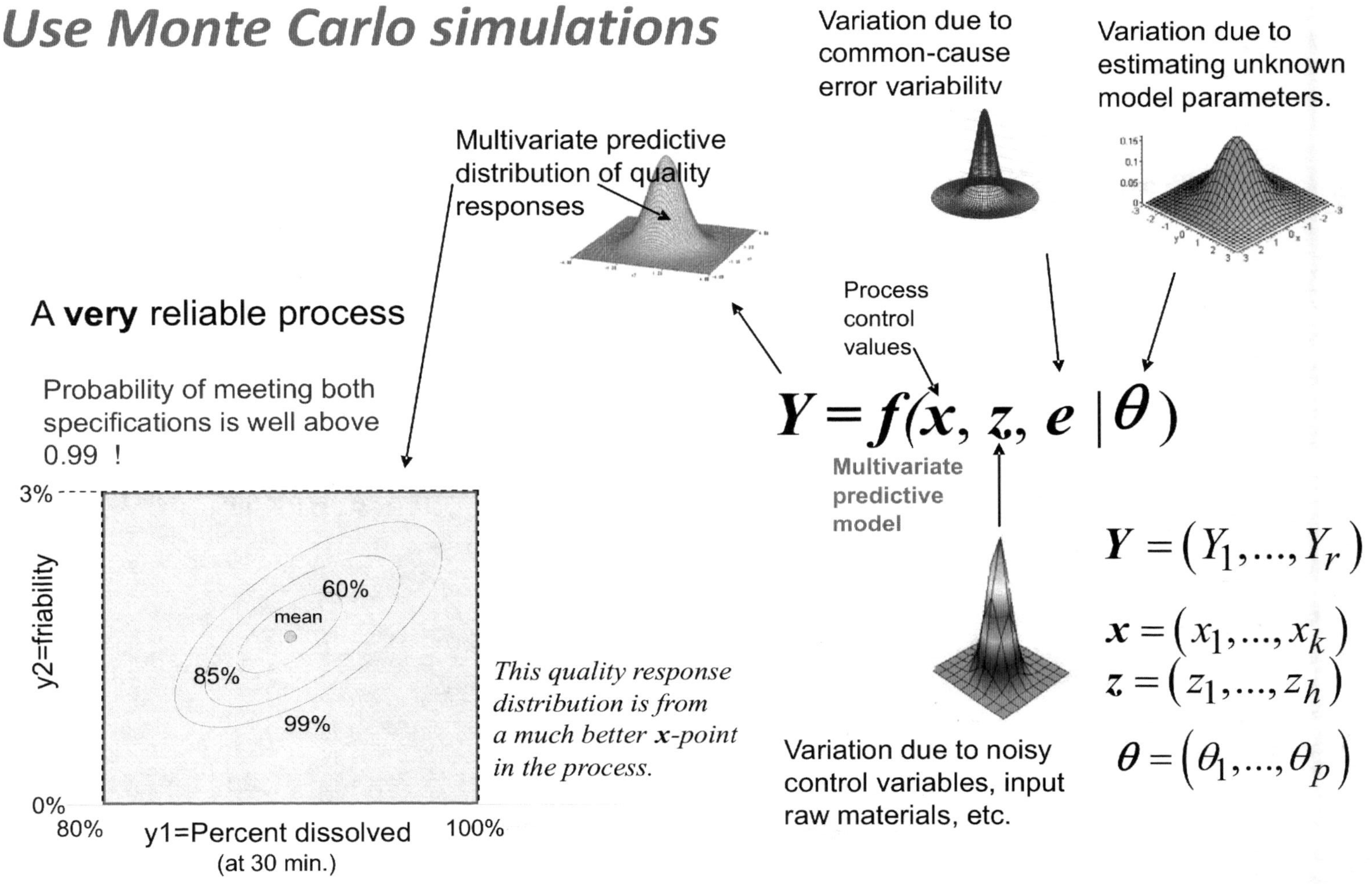

How do we construct a predictive distribution for our process?

Use Monte Carlo simulations

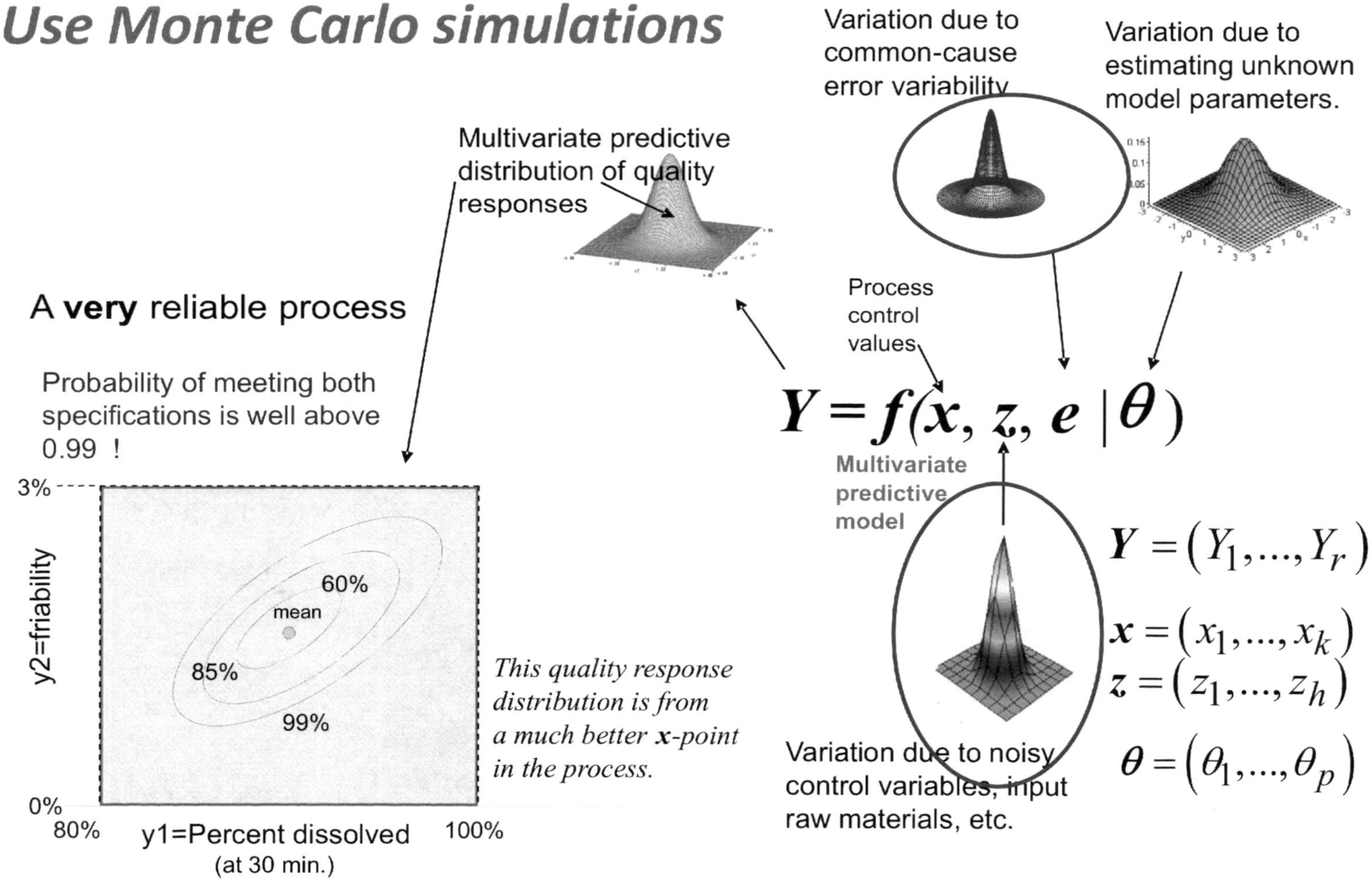

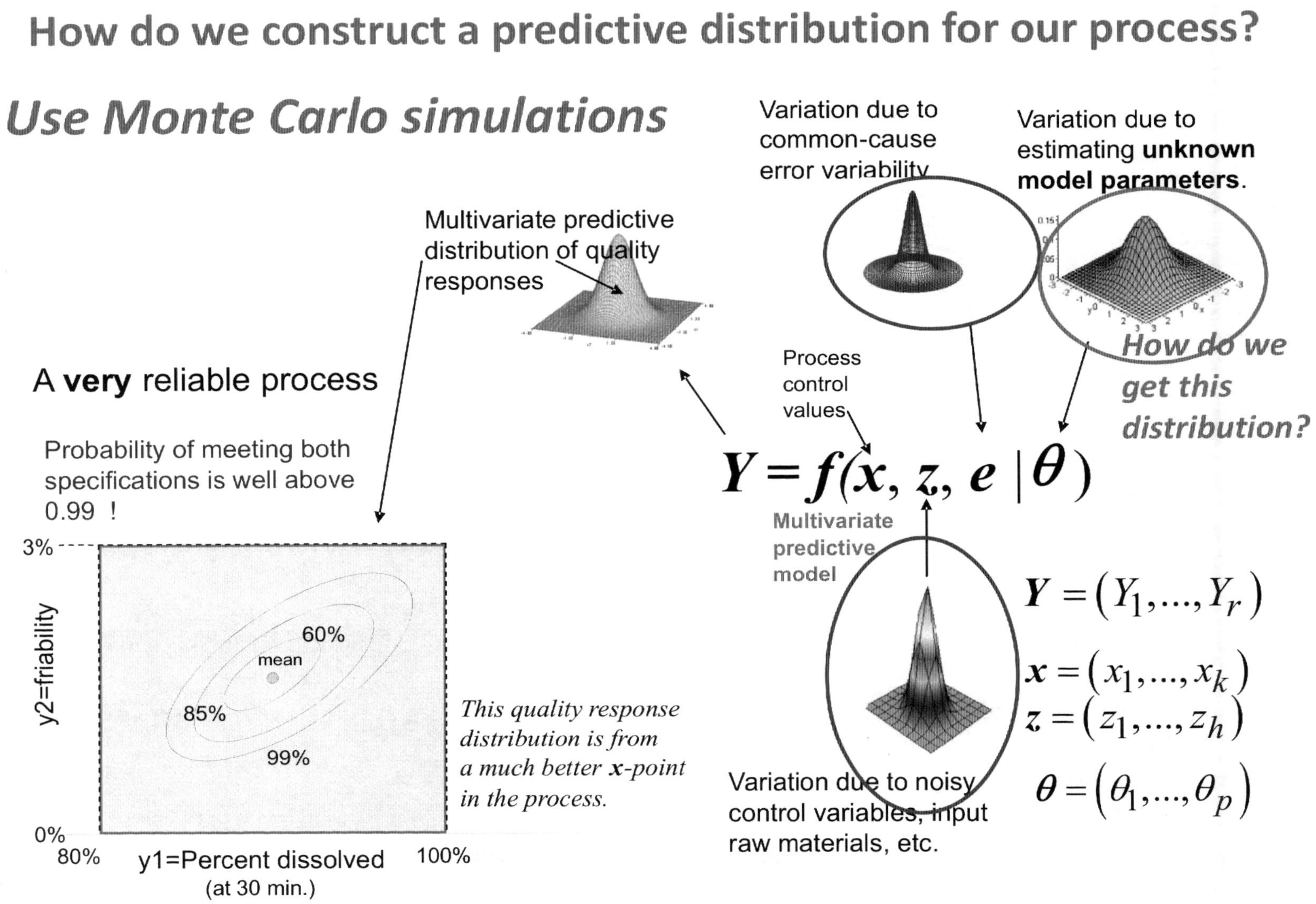
How do we construct a predictive distribution for our process?
Use Monte Carlo simulations
Variation due to common-cause error variability
Variation due to estimating **unknown model parameters**.
Multivariate predictive distribution of quality responses
How do we get this distribution?
A **very** reliable process
Process control values
Probability of meeting both specifications is well above 0.99 !
$Y = f(x, z, e \mid \theta)$
Multivariate predictive model
3%
0%
80%
100%
y2=friability
y1=Percent dissolved (at 30 min.)
mean
60%
85%
99%
This quality response distribution is from a much better x-point in the process.
$Y = (Y_1, ..., Y_r)$
$x = (x_1, ..., x_k)$
$z = (z_1, ..., z_h)$
$\theta = (\theta_1, ..., \theta_p)$
Variation due to noisy control variables, input raw materials, etc.

Can we assume a normal (i.e. Gaussian) distribution
for the unknown model parameters?

Answer: Sometimes yes, sometimes no.

Some possible distributions for two mechanistic model parameters:

Gaussian

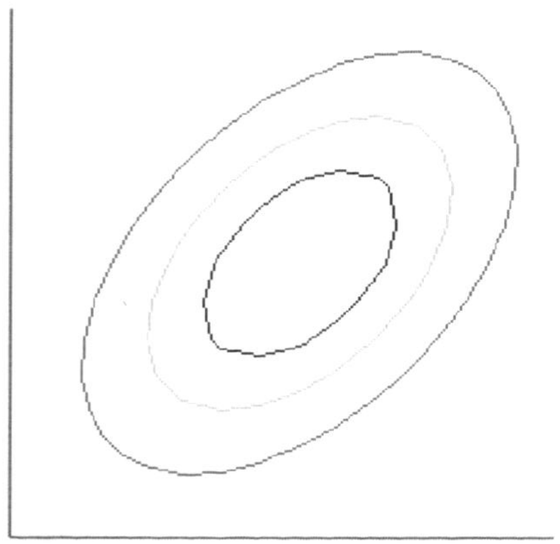

Skewed

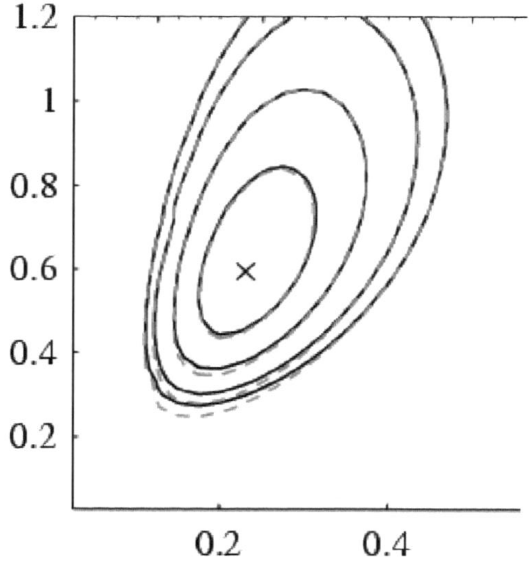

Banana-shaped

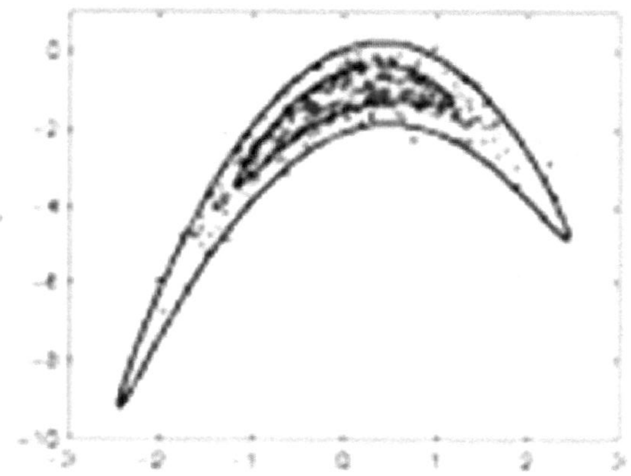

The Bayesian approach to obtain a multivariate distribution of unknown model parameters from experimental data

posterior distribution of unknown model parameters

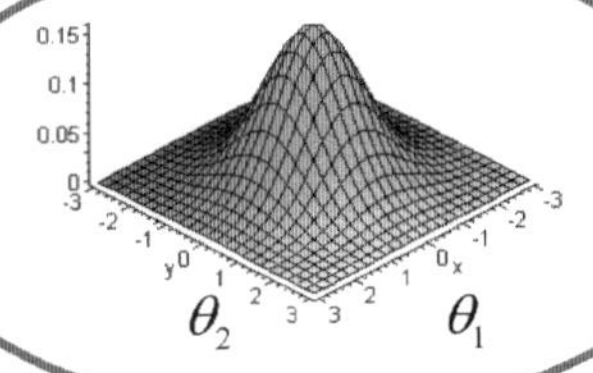

Bayes' rule weighting function

prior distribution of unknown model parameters

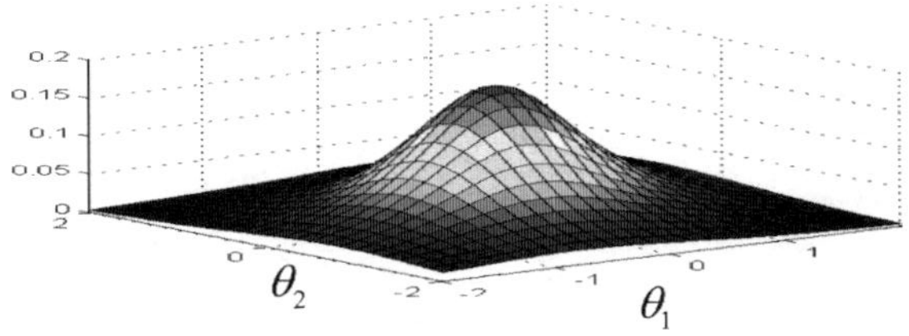

$$p(\theta \mid \text{data}) = \frac{L(data \mid \theta)}{\int L(data \mid \theta)\pi(\theta)d\theta} \times \pi(\theta)$$

- A probability procedure known as "Markov Chain Monte Carlo" (**MCMC**) can be used to obtain the posterior distribution of unknown model parameters given experimental data (and possibly prior information).

- It is possible to sample from the posterior distribution of the unknown model parameters. **Some Chem. E's are already doing this! See reference list for Blau et al. (2008) and Hsu et al. (2009)**

A "simple" parametric bootstrap approach to obtain a multivariate distribution of unknown model parameters from experimental data

1. Fit model to the data to obtain an estimate, $\hat{\boldsymbol{\theta}}$, of the vector of unknown model parameters, $\boldsymbol{\theta}$.

2. Use $\hat{\boldsymbol{\theta}}$ to simulate many new sets of "data": $\boldsymbol{Y}_1^{(1)},\ldots,\boldsymbol{Y}_n^{(1)}$ data set 1

$$\vdots$$

$$\boldsymbol{Y}_1^{(10,000)},\ldots,\boldsymbol{Y}_n^{(10,000)} \quad \text{data set 10,000}$$

3. Use <u>each</u> new data set to obtain a new parameter estimate: $\hat{\boldsymbol{\theta}}^{(1)},\ldots,\hat{\boldsymbol{\theta}}^{(10,000)}$

4. The set of parameter estimates in 3. above forms a distribution that reflects the uncertainty of the unknown model parameters.

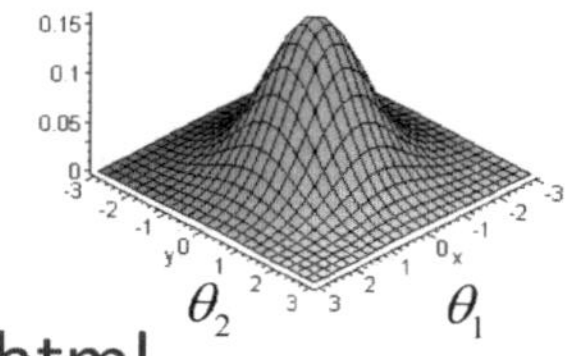

See http://www.pharmamanufacturing.com/articles/2010/097.html for details.

So...how can we improve quality and construct a reliable design space?

1. Build a *stochastic* model of your process.
   a. This can be a mechanistic or empirical model, as long as it predicts well.
   b. This stochastic model should produce a ***distribution*** of predicted values.

2. Through sequential (and designed) experimentation, find conditions to reduce process variation about the target.

3. The design space is the set of all process conditions (and inputs) that are associated with acceptably small variation (i.e. high reliability) about the target.

PS

Don't forget to model the uncertainty of your *unknown* model parameters!

Some Cautionary Notes about Multivariate Design Space and Process Modeling

- As far as I know, **no** point-n-click software package produces a multivariate predictive distribution for design space construction.

- Some packages may compute prediction intervals for each single response type, but they ignore the correlation structure of the process.

- The probability of meeting all specifications simultaneously will depend on the correlation structure of the multivariate predictive distribution. This dependence increases with the number of responses.

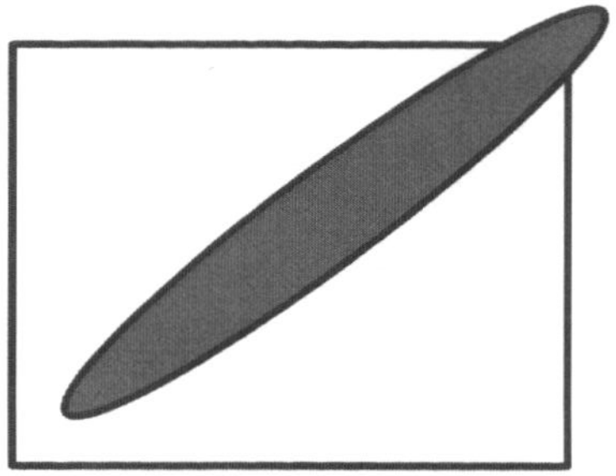

90% within specifications for positive correlation

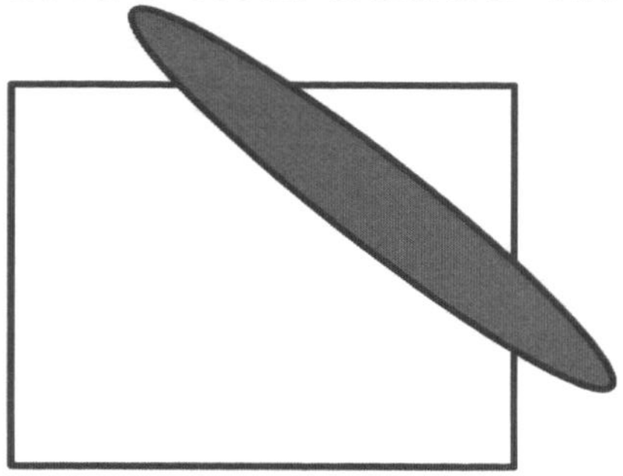

80% within specifications for negative correlation

Design Space Example 2: Design space can be determined from the common region of successful operating ranges for multiple CQA's. The relations of two CQA's, i.e., friability and dissolution, to two parameters are shown in Figures 2a and 2b. Figure 2c shows the overlap of these regions and the maximum ranges of the potential design space.

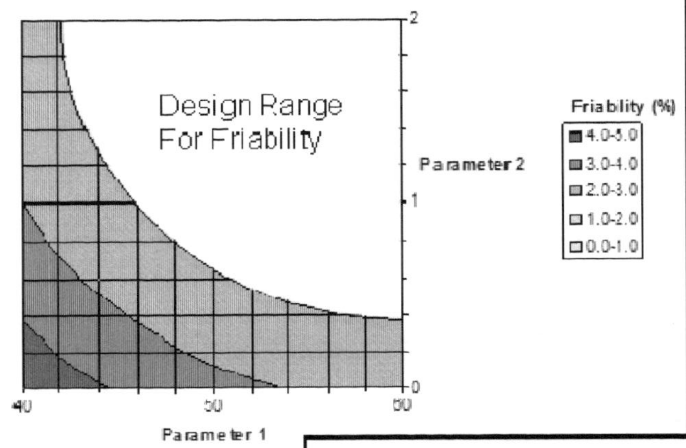

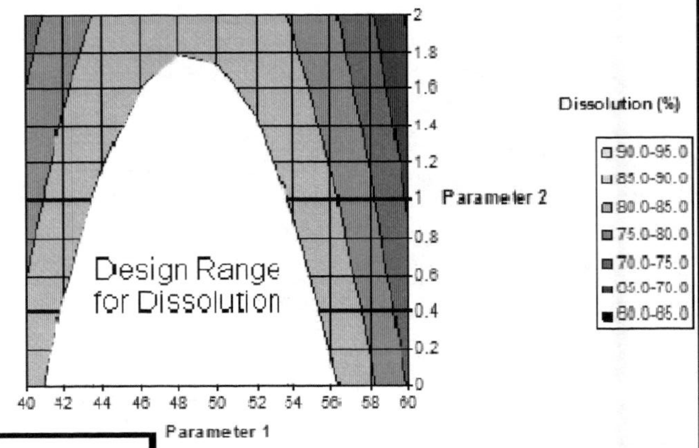

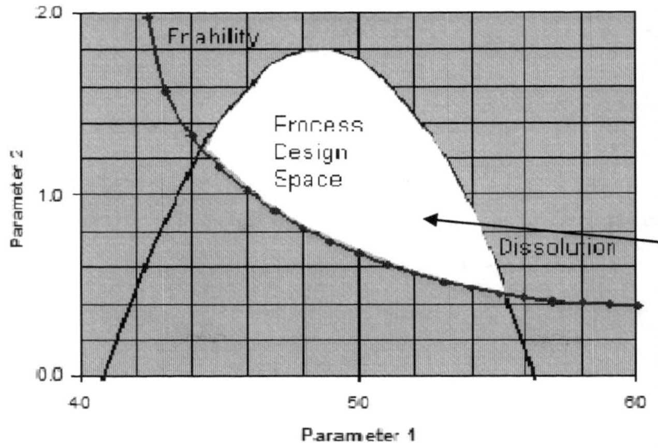

Taken from the ICH Q8 (Revised) (June 2009)

What do these contours represent? Mean response surfaces?

This overlay plot does not quantify "how much assurance"!

Overlapping Means vs. Bayesian Reliability Approach to Design Space:

An Example – due to Greg Stockdale, GSK.

Example: An intermediate stage of a multi-stage route of manufacture for an Active Pharmaceutical Ingredient (API).

Measurements:

- Four controllable quality factors (x's) were used in a designed experiment.
  (x1='catalyst', x2= 'temperature', x3='pressure', x4='run time'.)

- A (face centered) Central Composite Design (CCD) was employed.
  (It was a Full Factorial (30 runs), with no aliasing.)

- Four quality-related response variables, Y's, were measured.
  (These were three side products and purity measure for the final API.)

Y1= 'Starting material Isomer', Y2='Product Isomer', Y3='Impurity #1 Level', Y4='Overall Purity measure'

- Quality Specification limits: Y1<=0.15%, Y2<=2%, Y3<=3.5%, Y4>=95%.

Multidimensional Acceptance region, $A = [0, 0.0015] \times [0, 0.02] \times [0, 0.035] \times [0.95, 1]$

Overlapping Means vs. Bayesian Reliability Approach to Design Space:

An Example – due to Greg Stockdale, GSK.

Prediction Models:	Model Terms														
Response	x1	x2	x3	x4	x11	x22	x33	x44	x12	x13	x14	x23	x24	x34	
SM Isomer	Δ	Δ	Δ	Δ		Δ	Δ	Δ				Δ			
Prod Isomer	Δ		Δ	Δ			Δ			Δ					
Impurity	Δ		Δ							Δ					
Purity	Δ		Δ		Δ										

Temperature = x1 Pressure = x2 Catalyst Amount = x3 Reaction time = x4

Design Space Table of Computed Reliabilities[1] for the API (sorted by joint probability)

Note that the largest probability of meeting specifications is only about 0.75

Temp	Pressure	Catalyst	Rxntime	Joint Prob	SM Isomer Prob	Prod Isomer Prob	Impurity Prob	Purity Prob
35	60	6	3	0.752	1	0.9985	0.8435	0.79
32.5	60	7	3	0.743	1	0.9995	0.7875	0.8295
[illegible]	[illegible]	[illegible]	3	0.7375	0.9995	0.9995	0.7855	0.8255
[illegible]	[illegible]	[illegible]	3	0.737	1	0.9975	0.821	0.7845
[illegible]	[illegible]	[illegible]	3	0.7335	1	0.9995	0.7775	0.8175
37.5	60	6.5	3	0.725	1	1	0.7485	0.845
35	60	6.5	3	0.7225	1	1	0.77	0.812
32.5	60	6	3	0.7195	1	0.9955	0.864	0.7415
30	60	7	3	0.717	1	0.999	0.8075	0.759
32.5	60	7.5	3	0.716	1	1	0.734	0.859
37.5	60	5.5	3	0.7145	1	0.993	0.8065	0.7565
35	60	7	3	0.712	1	1	0.731	0.8555

Optimal Reaction Conditions

Marginal Probabilities

[1] This is only a small portion of the Monte Carlo output.

Overlapping <u>Mean</u> Contours from analysis of each response individually.

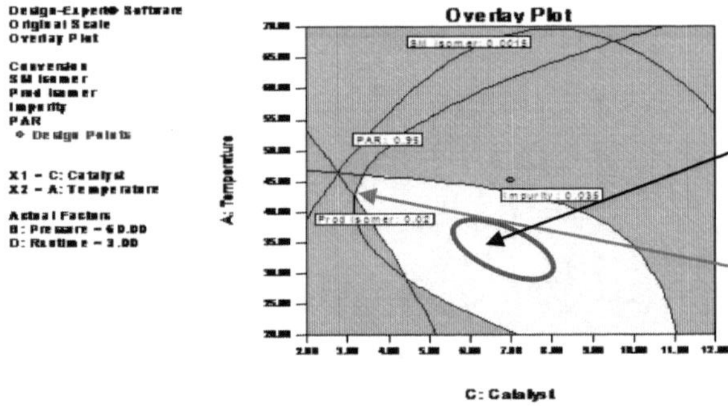

This x-point (in the yellow sweet spot) has only a probability of **0.75 .**

But this x-point (in the yellow sweet spot) has a probability of only 0.23 !

Posterior Predicted Reliability with

Temp=20 to 70, Catalyst=2 to 12, Pressure=60, Rxntime=3.0

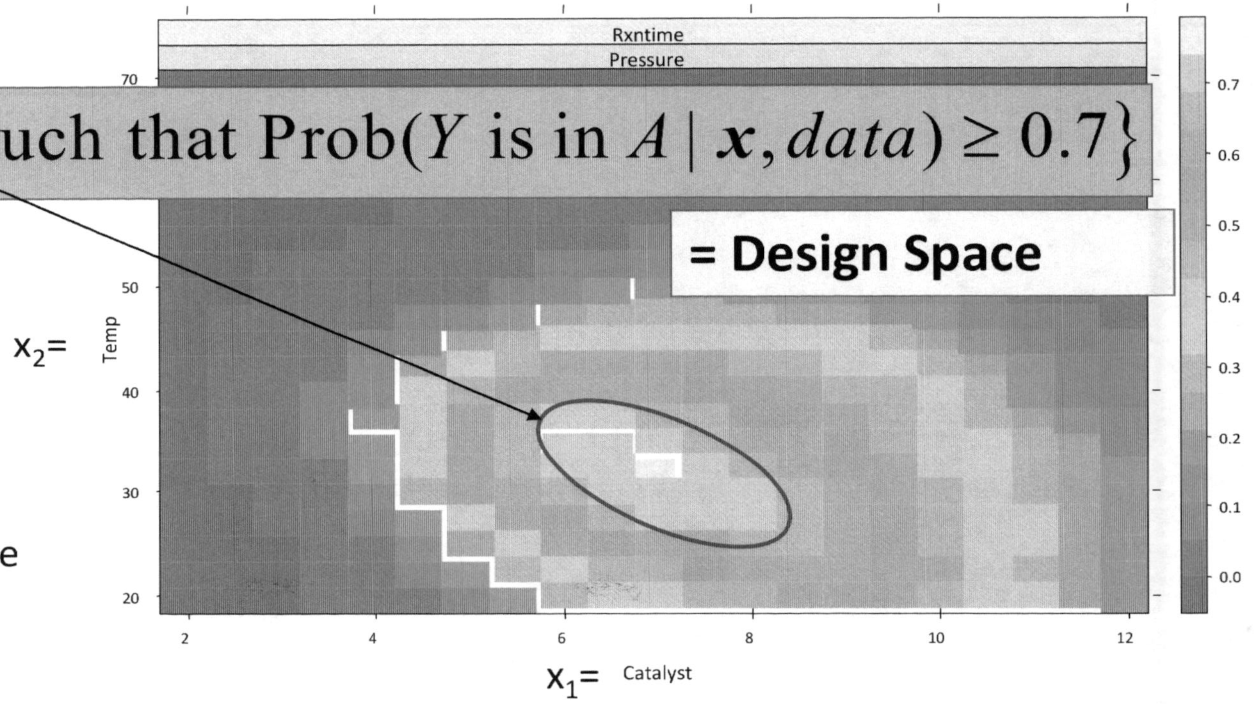

$$\{\boldsymbol{x} \text{ such that } \text{Prob}(Y \text{ is in } A \mid \boldsymbol{x}, data) \geq 0.7\}$$

= Design Space

Contour plot of $p(\boldsymbol{x})$ equal to Prob ($\boldsymbol{Y}$ is in A given $\boldsymbol{x}$ & data).

The region inside the red ellipse is the design space.

How to move optimally within a multi-dimensional design space

1. (Read-only) Sortable spreadsheet approach:

(i) Store the entire design space in a (read-only) sortable spreadsheet.

(ii) Each row should correspond to a point in the design space **and** the associated probabilities of meeting quality specifications.

(iii) If one factor (e.g. temp.) needs to be changed from 70°C to 80°C, then one can sort on other factors to see what configurations are associated with the best probability of meeting specifications.

See Peterson (2008) for details.

2. Meta-model approach:

(i) Using the data in the spreadsheet above, fit a flexible model (e.g. spline-based model) to obtain a (closed-form) approximating model for the design space probabilities.

(ii) Use mathematical optimization algorithms (applied to the approximating model) to move optimally within the multi-dimensional design space.

References

Blau, G., Lasinski, M., Orcun, S., Hsu, S-h, , Caruthers, J., Delgass, N., and Venkatasubramanian, V., (2008) “High fidelity mathematical model building with experimental data: A Bayesian approach”, *Computers and Chemical Engineering* 32, 971–989

del Castillo, E. (2007), *Process Optimization - A Statistical Approach*, Springer, New York, NY.

Hsu, S-H, Stamatis, S.D., Caruthers, J.M., Delgass, N.W., Venkatasubramanian, V., Blau, G., Lasinski, M. and Orcun, S. (2009), “Bayesian Framework for Building Kinetic Models of Catalytic Systems”, *Ind. Eng. Chem. Res*. 48, 4768–4790

Kenett, R. S. (2009), “By Design”, *Six Sigma Forum Magazine*, Nov. issue, pp27-29.

Miro-Quesada, G., del Castillo, E., and Peterson, J.J., (2004) "A Bayesian Approach for Multiple Response Surface Optimization in the Presence of Noise Variables", *Journal of Applied Statistics*, 31, 251-270

Peterson, J. J. (2004), "A Posterior Predictive Approach to Multiple Response Surface Optimization”, *Journal of Quality Technology*, 36:139-153.

Peterson, J. J. (2008), “A Bayesian Approach to the ICH Q8 Definition of Design Space”, *Journal of Biopharmaceutical Statistics*, vol. 18, pp959-975.

References (continued)

Peterson, J. J. (2009), "What Your ICH Q8 Design Space Needs: A Multivariate Predictive Distribution", *Pharmaceutical Manufacturing*, Nov./Dec. issue, pp23-28. available at: http://www.pharmamanufacturing.com/articles/2010/097.html

Peterson, J. J. and Yahyah, M., (2009) "A Bayesian Design Space Approach to Robustness and System Suitability for Pharmaceutical Assays and Other Processes", *Statistics in Biopharmaceutical Research* **1**(4), 441-449.

Peterson, J. J. Snee, R. D., McAllister, P.R., Schofield, T. L., and Carella, A. J., (2009) "Statistics in the Pharmaceutical Development and Manufacturing" (with discussion), *Journal of Quality Technology*, 41, 111-147.

Peterson, J. J. and Lief, K. (2010), "The ICH Q8 Definition of Design Space: A Comparison of the Overlapping Means and the Bayesian Predictive Approaches", *Statistics in Biopharmaceutical Research*, 2, 249-259.

Savage, S. (2009) *The Flaw of Averages - Why We Underestimate Risk in the Face of Uncertainty*, John Wiley and Sons, Inc., Hoboken, NJ.

Stockdale, G. and Cheng, A. (2009), "Finding Design Space and a Reliable Operating Region using a Multivariate Bayesian Approach with Experimental Design", *Quality Technology and Quantitative Management*, 6(4), 391-408

- **Two Take-Home Questions:**

  1. If someone shows you a real design space, ask: "**How much assurance** is there of meeting quality specifications for the worst point in that design space?"

  2. Does your design space take into account the uncertainty of all of the unknown model parameters, *and* the correlation structure of the process?

In Summary:

The mathematical and probabilistic methods exist to do the proper computations for ICH Q8 design space, but much technical modeling and programming work still needs to be done!

(358b) Optimization-Based Design Space Characterization Using First-Principles Models

Constantinos C. Pantelides[1], Mark Pinto[2] and Sean K. Bermingham[2], (1)Process Systems Enterprise Ltd., London, United Kingdom, (2)Strategic Business Development - Solids, Process Systems Enterprise Ltd., London, United Kingdom

Placeholder for Pantelides talk

See more of this Session: QbD Topical Plenary Session
See more of this Group/Topical: Topical i : Comprehensive Quality by Design in Pharmaceutical Development and Manufacture

(358c) Accelerated Development and Improved Process Operations Using Advanced Multivariate Latent Variable Modeling

Salvador Garcia-Munoz, Process Modeling & Engineering Technology, Pfizer Global Research & Development, Groton, CT

This talk will focus on the use of advanced multivariate latent variable models to aid the accelerated development of the product and the process, as well as the incorporation of these models into a process improvement cycle at the manufacturing scale.

See more of this Session: QbD Topical Plenary Session
See more of this Group/Topical: Topical i : Comprehensive Quality by Design in Pharmaceutical Development and Manufacture

(358d) A Perspective on Batch Versus Continuous Processing

Christine Seymour, Pfizer Global Research & Development, Pfizer, Groton, CT

placeholder for Chris' talk

See more of this Session: QbD Topical Plenary Session
See more of this Group/Topical: Topical i : Comprehensive Quality by Design in Pharmaceutical Development and Manufacture

(358e) Quality by Design and Multivariate Process Representations

Jose Cardoso Menezes, Chemical & Biological Engineering Department, I.S.T., Lisbon, Portugal

Bioprocess design, development and manufacturing involve different goals and use different tools. At the design stage a cell-centric approach dominates. During development and especially later during manufacturing a process-centered perspective prevails as if biomanufacturing performance was dictated by process variables alone. At the end of downstream processing a product is isolated in required purity and only then a link is established to the desired targeted product profile.

A stronger link between process and product must be established throughout all those stages in order to bring biomanufacturing into the XXI century and significantly improve process performance, reduce process variability and moreover ensure consistent product quality, efficacy and safety at the end. Process analytical technology represents the combined use of different tools applicable through many of those stages. Though PAT is still mostly process-centered it can be used within the QbD (quality by design) context to link process to product.

A study of a bacterial cultivation process for the production of a secondary metabolite was made at several scales (lab to plant) using multivariate design space concepts. A multivariate process representation is more informative of process changes along batch processing time and also across different scales, than descriptions based on univariate formulations of culture or process parameters. Univariate classical approaches can thus be replaced by more straightforward, summarized and reliable process state representations based on whole process analysis.

In our talk we will show the equivalency of using both approaches by using in-situ FT-NIR spectroscopy to fingerprint the cultivation media and in doing so map the process trajectories throughout a batch. The proposed approach is fully in-line with the definition of a process design-space containing the information needed for accurate scale-up and process optimization.

See more of this Session: QbD Topical Plenary Session
See more of this Group/Topical: Topical i : Comprehensive Quality by Design in Pharmaceutical Development and Manufacture

(424a) A Quality Systems Approach to the Product Quality Risk Management Lifecycle

Steven Wolfgang, Office of Compliance, Division of Manufacturing and Product Quality, FDA Center for Drug Evaluation and Research, White Oak, MD

As the knowledge base is built during development many of the risks to product quality and patient safety are identified and linked to inputs involving raw materials, equipment and facility design and processes. A comprehensive understanding about how all of these variables influence one another and interact allows manufacturers to implement sophisticated process control strategies and technologies that minimize process and product variability. A sound approach to quality risk management also utilizes the knowledge and experience gained after commercialization from monitoring of product quality and process performance. High quality products and reliable processes are maintained throughout the product lifecycle via the quality system infrastructure which collects and manages new information and relates it to the existing knowledge base. This presentation will discuss how quality risk management continues beyond product development, showing the vital role an effective quality system plays in maximizing the effectiveness of risk management in making continual improvements throughout the product lifecycle. ••

See more of this Session: Quality-by-Design (QbD): Techniques for Quality Risk Analysis and Management Over the Product Life Cycle
See more of this Group/Topical: Topical i : Comprehensive Quality by Design in Pharmaceutical Development and Manufacture

(424b) Implementation of a Normalization Risk Assessment Tool to Link Design Space Inputs

Diana S. Hou, Neil MacPhail, Edward J. Smith, Jennifer Ho and Alanna Cleary, Merck Manufacturing Division, Merck & Co., West Point, PA

Risk analysis and assessment tools are a key component of Quality by Design (QbD) because they facilitate understanding, identifying, and documenting of risks affecting drug product quality attributes. As stated in the ICH Q8 guidelines, the "design space" is where a process generates a product that meets all of it's critical quality attributes through a multidimensional combination and interaction of input variables and process parameters. The design space should be developed by addressing the risks identified in the risk analysis. There are several methods to prioritize the risk analysis to develop a mitigation strategy of the risks to bracket the desired design space.

Using a normalization tool, we have demonstrated a methodology to address the issue of sorting and prioritizing the risk analysis information. This tool enables standardization of the analysis of the risk assessment tool across development teams and projects. A case study for a high shear granulated product is presented to demonstrate how this tool can be used to identify key factors from the risk assessments and to drive the development work needed to support the definition of the design space.

See more of this Session: Quality-by-Design (QbD): Techniques for Quality Risk Analysis and Management Over the Product Life Cycle
See more of this Group/Topical: Topical i : Comprehensive Quality by Design in Pharmaceutical Development and Manufacture

(424c) Risk Based Roadmap for the Approval of Biosimilar Biologics

Henry Y. Wang, University of Michigan, Ann Arbor, MI

The development of biologics has grown significantly in recent years, due in part to the profitability for innovator companies, which currently enjoy exclusivity in the absence of "generic" biologics. In the U.S. There is currently no guideline in place for the approval of "biosimilar" biologics. With the rapid growth of biologics production and several biopharmaceutical patents nearing their expiration dates, a regulatory system is needed to classify "biosimilar" biologics in order to allow abbreviated FDA approval whenever possible. A risk based classification system has the potential to allow for a more streamlined, cost-effective approach for assessing biosimilarity and safety, which could lead to more affordable biologics for patients. We have proposed a new risk based Classification System for Biologics (CSB) to establish biosimilarity based on risk assessment and level of characterization of the product. Immunogenicity and stability testing will be required for determining the possibility of a biowaiver with or without additional clinical testing.

See more of this Session: Quality-by-Design (QbD): Techniques for Quality Risk Analysis and Management Over the Product Life Cycle
See more of this Group/Topical: Topical i : Comprehensive Quality by Design in Pharmaceutical Development and Manufacture

(424d) An Example of Quality Risk Management Over Product Lifecycle

T. Kourti, McMaster University, GlaxoSmithKline, Hamilton, ON, Canada

An example from a tablet product will be presented where quality for the product lifecycle is monitored in a "soft sensor" sense. A multivariate model derived from process parameters across many unit operations as well as API properties is shown to provide predictions on quality as well as alerts for possible issues.

See more of this Session: Quality-by-Design (QbD): Techniques for Quality Risk Analysis and Management Over the Product Life Cycle
See more of this Group/Topical: Topical i : Comprehensive Quality by Design in Pharmaceutical Development and Manufacture

(424e) Quality by Design in Generic Drug Development: Process Development and Quality Risk Management

Sivakumar R. Vaithiyalingam, Division of Chemistry III, Office of Generic Drugs, Center for Drug Evaluation and Research, FDA, Rockville, MD

The recent ICH guidelines Q8 (R2)(Pharmaceutical Development), Q9 (Quality Risk Management), and Q10 (Pharmaceutical Quality System) outline a comprehensive Quality-by-Design (QbD) approach. For the QbD process development, among the critical elements highlighted in the ICH Q8(R2), the ability of the process to reliably manufacture a product of the intended quality with respect to its clinical performance needs to be assessed. Process development is a systematic evaluation that includes prior knowledge, experimentation, and risk assessment. Such systematic and enhanced product and process understanding in combination with quality risk management could establish an appropriate control strategy which can, for example, include continuous process monitoring and real time feed back controls.

Risk management is integral part of QbD. Evaluation of risk to quality should be based on scientific knowledge and ultimately linked to the safety, efficacy and quality of the drug product to the patient. The level of effort, formality and documentation of the quality risk management process should be commensurate with the level of risk itself.

An effective quality risk management approach can further ensure the high quality of the drug (medicinal) product to the patient by providing a proactive means to identify and control potential quality issues during development and manufacturing.

This presentation will cover the above features of process development and quality risk management under QbD paradigm along with a case study for illustration.

See more of this Session: Quality-by-Design (QbD): Techniques for Quality Risk Analysis and Management Over the Product Life Cycle
See more of this Group/Topical: Topical i : Comprehensive Quality by Design in Pharmaceutical Development and Manufacture

(424f) A High Throughput, Parallel, Microscale, Fully Automated Approach to Biologics Formulation Development and Stress Test Studies

Eric Carlson[1], Byeong Chang[2], Steve Cypes[1] and Steve Lambert[1], (1)Freeslate Inc, Sunnyvale, CA, (2) Integrity, Bio Inc., Camarillo, CA

Developing robust formulations for biological materials (proteins, antibodies, vaccines, etc.) requires the screening and analysis of a comprehensive matrix of formulation compositions and stress conditions. This talk will focus on the development and use of a fully-automated workflow for stressing these types of formulations in their final container closure systems; such a system is intended to fit into a Quality by Design(QbD) program in which scientists seek "a full understanding of how product attributes and process relate to product performance based on scientific,risk-based, holistic and proactive approaches to product development." Guzman, R. "A Regulatory Approach to Contract Manufacturing", Pharmaceutical Outsourcing, March/April 2010, p 16-20. The system comprises automated environmental chambers (-20, 4, 25, and 40 deg C), several stationary sample preparation and measurement robots (Freeslate Core Module with pH probes, dispensing and mixing tools, balance and camera) and a multitude of analytical instruments (HPLCs, osmometer, electrophoresis-on-chip, UV-Vis/ fluorescence and dynamic light scattering plate readers) and it enables automated post-stress analytical sample workup and measurement. The system is integrated through an informatics system using Freeslates's Laboratory Execution and Analysis software to create an informatics platform that allows scientists to view, aggregate and make conclusions on large analytical datasets collected by the system. In addition, these instruments are physically integrated with a 6-axix robotic arm on a linear track, with changeable end-effectors, that moves sample plates, reagents and consumables to the various workstations. Data generated using the system for two case studies will be presented. The first study examines the effect of surfactant composition and concentration on protein stability. The second study examines the temperature stability of an antibody formulation as a function of pH and composition.

See more of this Session: Quality-by-Design (QbD): Techniques for Quality Risk Analysis and Management Over the Product Life Cycle
See more of this Group/Topical: Topical i : Comprehensive Quality by Design in Pharmaceutical Development and Manufacture

(424g) CFD Simulations for Scale up of Wet Milling in High Shear Mixers

Meng Yang and Richard V. Calabrese, Chemical & Biomolecular Engineering, University of Maryland, College Park, MD

Rotor-stator mixers are widely used in the chemical and pharmaceutical process industries for a variety of emulsion and dispersion applications, that must be scaled to geometrically or dynamically different equipment on the pilot and plant scale. When the process fluid dynamics plays a significant role, a reasonable approach is to simulate the bench, pilot scale and plant scale flow field and deformation rate phenomena and match these on all scales to extrapolate bench scale data with respect to the number of mill head passes and shear rate history.

To date, we have been simulating the flow fields in Silversion L4R (bench scale), 450LS (pilot scale) and 600LS (plant scale) in-line mixers. The working fluid is water and the flow is turbulent. We use a RANS formalism with a k - epsilon turbulence model and non-equilibrium wall functions. Due to the importance of the rotor-stator interaction, a sliding mesh CFD technique must be employed.

The velocity fields and turbulence fields inside these mixers will be compared and discussed along with pumping capacity and power number predictions. The deformation rate fields are also analyzed and compared. The implication of the results to scale-up of wet milling of pharmaceuticals will be also discussed by reference to bench scale data for crystal size reduction.

See more of this Session: Quality-by-Design (QbD): Techniques for Quality Risk Analysis and Management Over the Product Life Cycle
See more of this Group/Topical: Topical i : Comprehensive Quality by Design in Pharmaceutical Development and Manufacture

(424h) Optimizing and Controlling High Value Processes Using the Principles of Quality-by-Design (QbD) Advanced Tools Such as Real-Time in-Situ Particle Characterization, in-Situ Mid-Infrared Spectrometry, and Reaction Calorimetry

Leen Schellekens, Mettler-Toledo AutoChem, Inc., Columbia, MD, Des O'Grady, Particle Systems Characterization, Mettler-Toledo AutoChem, Inc., Columbia, MD and Jeffrey Sherman, Buisness Development and Advanced Technology Development, Mettler-Toledo AutoChem, Inc., Columbia, MD

Quality by Design (QbD) is a concept applied to gain true process understanding utilizing tools such as Design of Experiment (DoE), Risk Management, and Process Analytical Technology (PAT). Utilization of these principles and tools not only allow the chemist, engineer, or formulator to optimize a high value chemical or physical process, but also allows them to determine the robustness of a process. The study of the process robustness not only determines process failure conditions by testing the boundary conditions for operating parameters, but also determines acceptable ranges of operation without impacting the final product quality.

Several industrial examples are presented that demonstrate the importance of the ability to precisely control reactions and processes on a lab scale (10mL-20L) while emulating plant conditions. Additionally, appropriate PAT tools are utilized to monitor continuously how the process is performing under actual reaction conditions delivers critical insight regarding the process so that modifications may be made to deliver a robust, optimized process. For the industrial studies presented in this paper, real-time in-situ particle characterization tools, in-situ mid-infrared spectrometry, and reaction calorimetry have been utilized to deliver a deeper understanding of each process.

This paper includes highlights of several actual industrial studies performed at the lab or pilot plant scale. The results of these studies, either Scale-Up or Scale-Down, were implemented into actual manufacturing environments and demonstrate the value derived from the appropriate use of QbD Principles and PAT tools. Examples presented demonstrate application not only for high value chemical processes but also in unit operations such as crystallization, high shear wet milling, and granulation.

See more of this Session: Quality-by-Design (QbD): Techniques for Quality Risk Analysis and Management Over the Product Life Cycle
See more of this Group/Topical: Topical i : Comprehensive Quality by Design in Pharmaceutical Development and Manufacture

(445a) Optimal Material Selection for Oral Dosage Development and Manufacture Using Optimization Techniques and Multivariate Latent Variable Models

Salvador Garcia-Munoz[1], Jose Mercado[2], Israel Cotto[2], Victor Ruiz[2], Ivelisse Colon-Rivera[1], Denise Rivkees[3], Josh Shockey[4], Matt Shaffer[4] and Leah Appel[4], (1)Pfizer Global Research & Development, Pfizer Inc., Groton, CT, (2)Pfizer Global Manufacturing - Puerto Rico, Pfizer Inc., Vega Baja, PR, (3) Pfizer Global Manufacturing - Global Manufacturing Services, Pfizer Inc., Morris Plains, NJ, (4) Greenridge Consulting, Bend, OR

Multivariate latent variable models (such as Principal Components Analysis –PCA, and Projection to Latent Structures –PLS) have been reported in literature to transform a collection of physical properties measured for multiple materials into scores (or principal properties) with the ultimate objective of aiding a selection or screening process, which is ultimately done by the practitioner.

Even when using principal properties, it is not uncommon to find situations where there are more than two principal components (and hence principal properties) and the decision of which materials to use for an experimental design is not obvious and will require an exhaustive manual exercise to select materials with the optimal position in the score space such that the covariance of the principal properties of the selected materials is kept orthogonal and centered around the mean to avoid bias.

This work presents the application of multivariate methods coupled with mixed integer non-linear optimization methods to aid the selection process from a finite population of materials when the objective is to maximize the balanced and orthogonal information from the materials onto an experimental design. An objective function is formulated such that the off diagonals of the variance-covariance matrix of the sub-set of scores for the selected materials is minimized, while maximizing the main diagonal of the variance-covariance matrix while keeping the "design" centered around the mean to avoid bias.

Furthermore, when the properties of the materials are considered as a mixture, modifications to the PLS algorithm have been proposed to model the covariance structure of such data and aid the decision making process that involves the selection of materials from a given pool. In this work we also present two examples of such exercise: one for process development and one for commercial manufacture in support of a process improvement effort. For the process development example a formulation for a wet granulated tablet is optimized in order to maximize the slope of the hardness compression profile. The commercial manufacture example presents the case where historical data for the physical chemical properties of the different lots of raw materials were regressed to the final dissolution of the product, and the model is then used to determine which of the available lots in inventory are the optimal to be blended to minimize the variability in the quality of the final product.

See more of this Session: Applications of Multivariate Data Analysis in Pharmaceutical Process Development
See more of this Group/Topical: Topical i : Comprehensive Quality by Design in Pharmaceutical Development and Manufacture

(445b) Applications of Genetic Algorithms for Variable Selection in Pharmaceutical Development

Jose E. Tabora, Jacob Albrecht, James Bergum, Amanda Rogers and Victor Rosso, Bristol-Myers Squibb Company, New Brunswick, NJ

Genetic algorithms GA have been demonstrated to be powerful and efficient tools in the optimization of variable selection for empirical model generation [1]. In pharmaceutical process development the problem of variable selection is encountered in two distinct applications. The application of PAT technology generally requires the development of a quantitative relationship between analyte concentration and a multi-wavelength spectral response. The determination of which wavelengths should be selected for an optimal model is a challenging and time consuming task. Another area in which variable selection is a key optimization task is the development of surface response curves which requires careful selection of the factor terms that should be included to optimize model accuracy and predictive power. This presentation discusses detailed applications of genetic algorithms in the generation of partial least square models for PAT applications and surface response models for process risk assessment evaluation. Generally the solution from a genetic algorithm is a family of genomes of similar fitness from which the appropriate model is subsequently selected. This presentation will also address strategies for optimal subject selection from the final population representing the GA solution. The results will be compared to models obtained via careful statistical analysis of the factor-response relationships.

[1] Leardi, R. Journal of Chemometrics 15, 559-569 2001

See more of this Session: Applications of Multivariate Data Analysis in Pharmaceutical Process Development

See more of this Group/Topical: Topical i : Comprehensive Quality by Design in Pharmaceutical Development and Manufacture

(445c) Multivariate Analysis of Historical Process Data for Monitoring, Control and Scale-up

Samarth Rathore, David Reed and Marc Champagne, Analytical R&D, Eli Lilly, Indianapolis, IN

A quality by design (QbD) initiative requires a good foundation of organized data. Without this foundation, there could be no developments in use of prior knowledge or implementation of quality by design approach or in realizing full potential of Process Analytical Technology (PAT) work. Over viewing an historical process data presents many challenges including that of preprocessing of data, large amount of missing data, and large number of collinear measurements etc. Use of multivariate data analysis such as hierarchical PLS provides a useful tool for process understanding and continuous process improvement, and overcomes challenges faced in analyzing process data. Organization of data into hierarchical fashion not only facilitates data management, but also provides convenience in analyzing data. Present work gives examples from formulation, granulation, tablet press and analytical results data. Current work demonstrate use of multivariate methods, at the lower base level, to analysis local models for comparison of formulation, equipment setup and process parameters that provides a more detailed understanding at local level of the unit operations i.e. granulation and tablet press. Present work also demonstrate at the top level, how to relate quality of product with formulation, and process parameters and provide an overview of the entire process, and reflect the most dominant patterns and trends. Application of local models for investigating scale-up of unit operation, and use of global models for entire process transfer is discussed.

See more of this Session: Applications of Multivariate Data Analysis in Pharmaceutical Process Development
See more of this Group/Topical: Topical i : Comprehensive Quality by Design in Pharmaceutical Development and Manufacture

Estimation of Active Pharmaceutical Ingredients Content in Blending Process for Drug Products Manufacturing

Sanghong Kim*, Hiroshi Nakagawa†, Manabu Kano* and Shinji Hasebe*

Abstract

This study focuses on the content estimation of Active Pharmaceutical Ingredients (API) in the blending process from Near-Infrared spectrum (NIRS) data. Particle size data are used as inputs in addition to spectrum data to take account of the influence of particle size distribution on NIRS, and locally weighted partial least squares (LW-PLS) is applied to estimate API content. LW-PLS builds a local linear regression model based on the similarity between a query and sample data stored in the database when output estimation is required. LW-PLS can be adaptive and cope with nonlinearity in spite of its simplicity. In addition, a statistical wavelength selection method is proposed. This method can select a suitable set of wavelengths for the estimation; the selected wavelengths are found to include those corresponding to the spectrum peaks specific for API. LW-PLS and the proposed wavelength selection method were applied to real process data, and the estimation accuracy was improved by 18.7 % in Root Mean Square Error of Prediction (RMSEP) compared with the conventional PLS and wavelength selection method base on spectrum peak positions. The results clearly show that the proposed method is useful for the API content estimation and is superior to the conventional method.

1 Introduction

In the pharmaceutical industry, many researches have been conducted in order to improve production efficiency by developing on-line monitoring and control systems since PAT guidance was published in 2004. Especially, researches on drug quality estimation have been actively conducted and many kinds of qualities such as water content, blend homogeneity, content uniformity and coating thickness are estimated [1], [2]. However, little number of PAT are used in real production plants and there are still some problems to solve to make PAT widely used. First, it is difficult to construct a robust model which is applicable in real processes because physical and chemical properties of a measuring object have a complex effect on NIRS. In addition, optical scattering and autofluorescence can have non-linear effect on NIRS, therefore conventional linear PLS and PCR, which are used in many papers [3]~[11], seem not applicable in real processes in many cases. Second, it is difficult to choose a proper preprocessing method and wavelengths. Wavelength selection methods based on Genetic Algorithm [12] and latent variable vectors [13] were studied, however, the best way is different in every case [14]. Therefore, in the present situation, it is very laborious to select a proper preprocessing method and wavelengths, and it is required to develop a systematic and general method. Third, it is difficult to maintain model accuracy for a long period. Although model maintenance is required by the authorities, there are little papers focusing on this problem.

*Kyoto University, Dept. of Chemical Engineering

†Daiichi Sankyo Co., Ltd. Formulation Technology Research Laboratories

In this study, the method which can solve these problems is proposed and API content is estimated. To solve the first problem, particle size data are used as inputs in addition to spectrum data and LW-PLS is applied. LW-PLS can also solve the third problem because model adaptation is possible. To solve the second problem, a systematic wavelength selection method is proposed.

2 Locally weighted PLS algorithm

The nth measurements of input variables and API content value are denoted by $\boldsymbol{x}_n = [x_{n1}, x_{n2}, \cdots, x_{nM}]^{\mathrm{T}}$ $(n = 1, 2, \cdots, N)$ and y_n respectively, where M and N are the number of input variables and the number of samples, respectively. In addtion, x_{nm} $(m = 1, 2, \cdots, M)$ is the nth measurement of the mth input variable. $\boldsymbol{X} \in \Re^{N\times M}$ and $\boldsymbol{y} \in \Re^{N\times 1}$ are a matrix and a vector whose nth rows are $\boldsymbol{x}_n^{\mathrm{T}}$ and y_n, respectively. It is assumed here that each column of $\boldsymbol{X}$ is mean-centered. The locally weighted PLS (LW-PLS) algorithm is as follows.

1. Calculate the similarity

$$\omega_n \equiv \exp\left(-\frac{\| \boldsymbol{x}_q - \boldsymbol{x}_n \|}{\sigma_L \varphi}\right) \tag{1}$$

   between a query $\boldsymbol{x}_q$ and a measurement $\boldsymbol{x}_n$ $(n = 1, 2, \cdots, N)$. Here σ_L is standard deviation of $\| \boldsymbol{x}_q - \boldsymbol{x}_n \|$ and φ is a parameter which can control the effect of ω_n on the model. ω_n has large effect on the model when φ is small, but the model can easily be affected by noise when φ is too small.

2. Set $k = 1$ and $\hat{y}_q = 0$. Here $\hat{y}_q$ denotes an API content estimation value of the query.

3. Derive kth latent variable of $\boldsymbol{X}$, $\boldsymbol{t}_k = \boldsymbol{X}\boldsymbol{w}_k$, where $\boldsymbol{w}_k = \dfrac{\boldsymbol{X}^{\mathrm{T}}\boldsymbol{y}}{\| \boldsymbol{X}^{\mathrm{T}}\boldsymbol{y} \|}$.

4. Derive kth loading vector for $\boldsymbol{X}$, $\boldsymbol{p}_k = \dfrac{\boldsymbol{X}^{\mathrm{T}}\boldsymbol{t}_k}{\| \boldsymbol{t}_k^{\mathrm{T}}\boldsymbol{t}_k \|}$.

5. Derive kth coefficient for $\boldsymbol{y}$, $q_k = \dfrac{\boldsymbol{y}^{\mathrm{T}}\boldsymbol{\Omega}\boldsymbol{t}_k}{\| \boldsymbol{t}_k^{\mathrm{T}}\boldsymbol{\Omega}\boldsymbol{t}_k \|}$. Here $\boldsymbol{\Omega}$ is a diagonal matrix whose nth diagonal element is ω_n.

6. Derive kth latent variable of $\boldsymbol{x}_q$, $t_{qk} = \boldsymbol{x}_q\boldsymbol{w}_k$.

7. Set $\hat{y}_q = \hat{y}_q + t_{qk}q_k$.

8. If k equals to the predetermined latent variable number, finish estimation. Otherwise, set $k = k + 1$, $\boldsymbol{X} = \boldsymbol{X} - \boldsymbol{t}_k\boldsymbol{p}_k^{\mathrm{T}}$, $\boldsymbol{y} = \boldsymbol{y} - \boldsymbol{t}_k q_k$ and $\boldsymbol{x}_q = \boldsymbol{x}_q - t_{qk}\boldsymbol{p}_k^{\mathrm{T}}$ and return to step 3.

3 Wavelength selection method

In this section, the algorithm of the proposed wavelength selection method is explained. The proposed method supposes that there are data from different lots where API content is different. The wavelength selection procedure is as follows.

Table 1: Experimental data

Lot number	Number of samples	Average HPLC value [%]	Average particle size [μm]
1	100	88.7	145.1
2	20	97.4	152.5
3	90	107.7	134.1
4	90	68.1	184.6
5	86	83.0	159.7
6	10	98.6	158.8
7	90	113.8	140.4
8	90	122.9	133.7

1. Devide $\boldsymbol{X}$ into each lot data, then calculate mean value $\bar{x}_{*,m,l}$ and variance $\mathrm{V}_n(x_{n,m,l})$ in mth wavelength for lth lot. Here $x_{n,m,l}$ denotes the absorbance of thr nth sample in mth wavelength for lth lot.

2. Select wavelengths for model construction where $\dfrac{\mathrm{V}_l(\bar{x}_{*,m,l})}{\sum_l \mathrm{V}_n(x_{n,m,l})} \geq \lambda$ is satisfied. Here $\mathrm{V}_l(\bar{x}_{*,m,l})$ denotes the variance of $\bar{x}_{*,m,l}$ $(l = 1, 2, \cdots, L)$ where L is the number of lots.

Here λ is a threshold for variance ratio. By increasing λ, unnecessary information for API content estimation can be removed. However, too large λ leads to losing useful information. This method is based on the concept that $\mathrm{V}_n(x_{n,m,l})$, which means within-lot variance of absorbance (ideally zero), shoud be small and that $\mathrm{V}_l(\bar{x}_{*,m,l})$, which means sensitivity of absorbance to API content, should be large.

4 Application to real process

4.1 Experimental

The target drug formulation consists six components. In order to construct a model which can accurately estimate API content in wide range, eight blending experiments were carried out with different API content using 3 L scale V-blender (Tsutsui scientific instruments Co.,Ltd). After each blending experiment, the drug formulation was taken out and set in 200 mg vials, and NIRS (2203 points in 12493~4000 cm^{-1}) and API content were measured with MPA (Bruker Optics K.K.) and Alliance Waters 2690 Separations Module (Waters Corporation), respectively. In addition, average particle size in each lot was measured with ROBOT SIFTER RPS-85 (Seishin Enterprise Co, Ltd.). The overview of the experimental data is shown in Table 1.

4.2 Data analysis

In this study, data from lot 1~3 and 4~8 were used for model validation and construction, respectively. The procedure of the data analysis is as follows.

1. Preprocessing
   First order differential using Savitsky-Golay filter and Standard Normal Variate (SNV) were applied. By differentiating spectrum, the positions of peak become clear and SNV

can correct the variance in light path length caused by the change in the particle size and density [15].

2. Wavelength selection
   Here the proposed wavelength selection method and the conventional wavelength selection method were compared to check the usefulness of the proposed method. In the conventional method, at first, spectrum were secondary differentiated in order to identify the peak positions specific for API, then the optimal set of the identified peak wavelengths for estimation was determined using the stepwise method.

3. Model construction
   LW-PLS and conventional PLS were compared. When particle size data were used as inputs in addition to spectrum data, the variance of particle size was scaled to maximum, mean and median of the variance of the absorbance in selected wavelengths.

16 estimation models were constructed with respect to selections in step 2 and 3. Model parameters in each model, i.e. φ, λ, number of latent variables and window size of Savitky-Golay filter, were determined through cross validations.

4.3 Results and discussion

The applied methods and estimation results are shown in Table 2 and Figs. 1∼16. In Figs. 1 ∼16, r denotes the correlation coefficient between measured and estimated API content values. Table 2 shows that the proposed wavelength selection method is superior to conventional one, and RMSEP became worse when the particle size was used as an input variable. This may be because there is insufficient number of data (only one particle size data for each lot). LW-PLS achieved better RMSEP when it was applied with the proposed wavelength selection method while RMSEP became worse when LW-PLS was applied with the conventional wavelength selection method. This means that φ could not be determined properly through cross validations, because the estimation accuracy of LW-PLS is always better than or equal to that of PLS if φ is determined properly. φ may be determined properly if there are more experimental data. In addition, even if there is sufficient number of data, the similarity ω depends on the selected wavelengths, therefore the wavelengths which are suitable for expressing the similarity between API content of a query and sample data should be selected.

5 Conclusions

In this study, LW-PLS, a systematic wavelength selection method and utilizing the particle size data were proposed to develop a general drug quality estimation system and the proposed method was applied to real process data to estimate API content. The results show that LW-PLS and the proposed wavelength selection method are useful, although utilizing the particle size made the estimation accuracy worse.

Table 2: Estimation methods and results

Case	Wavelength selection	Modeling	Input variable	RMSEP
1	Conventional	PLS	Spectra	1.98
2	Conventional	PLS	Spectra + particle size (max)	2.46
3	Conventional	PLS	Spectra + particle size (mean)	2.46
4	Conventional	PLS	Spectra + particle size (median)	2.46
5	Conventional	LW-PLS	Spectra	2.01
6	Conventional	LW-PLS	Spectra + particle size (max)	2.50
7	Conventional	LW-PLS	Spectra + particle size (mean)	2.51
8	Conventional	LW-PLS	Spectra + particle size (median)	2.51
9	Proposed	PLS	Spectra	1.68
10	Proposed	PLS	Spectra + particle size (max)	2.03
11	Proposed	PLS	Spectra + particle size (mean)	1.98
12	Proposed	PLS	Spectra + particle size (median)	1.79
13	Proposed	LW-PLS	Spectra	1.61
14	Proposed	LW-PLS	Spectra + particle size (max)	1.95
15	Proposed	LW-PLS	Spectra + particle size (mean)	1.96
16	Proposed	LW-PLS	Spectra + particle size (median)	1.94

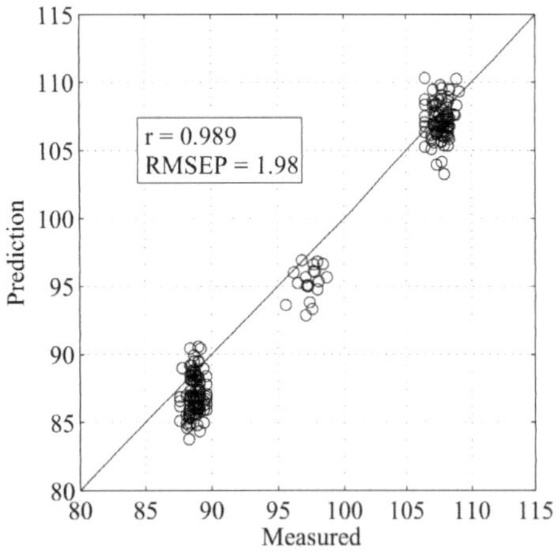

Figure 1: Estimation result (Case 1)

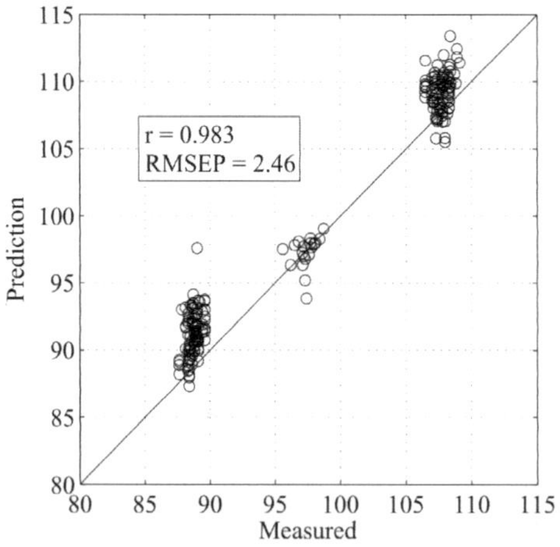

Figure 2: Estimation result (Case 2)

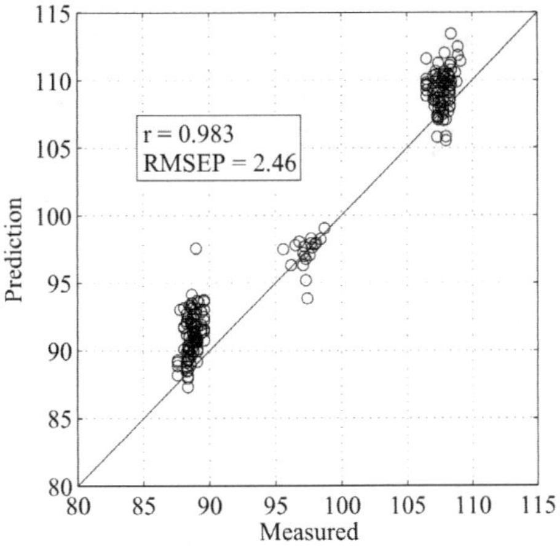

Figure 3: Estimation result (Case 3)

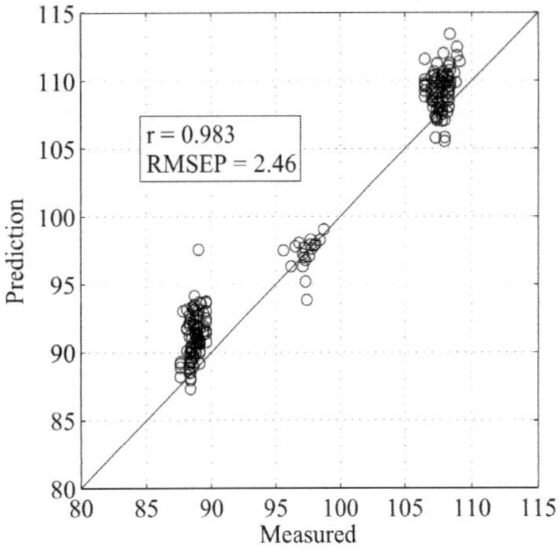

Figure 4: Estimation result (Case 4)

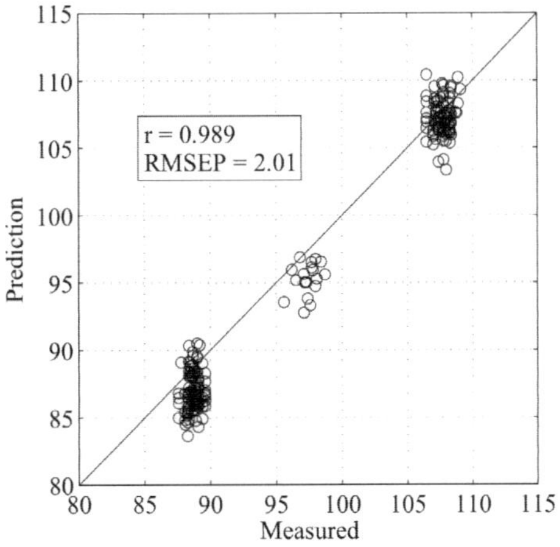

Figure 5: Estimation result (Case 5)

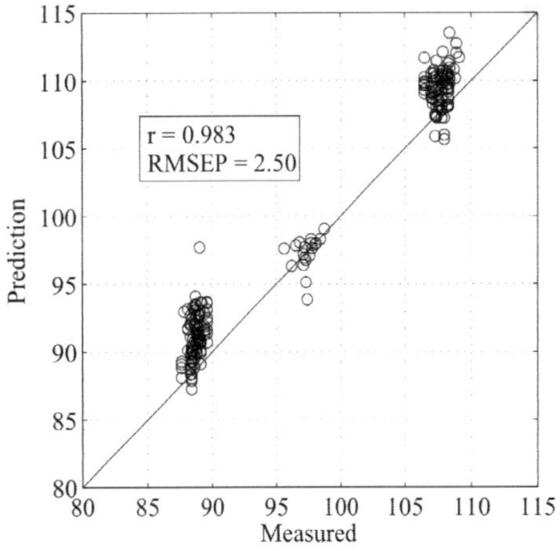

Figure 6: Estimation result (Case 6)

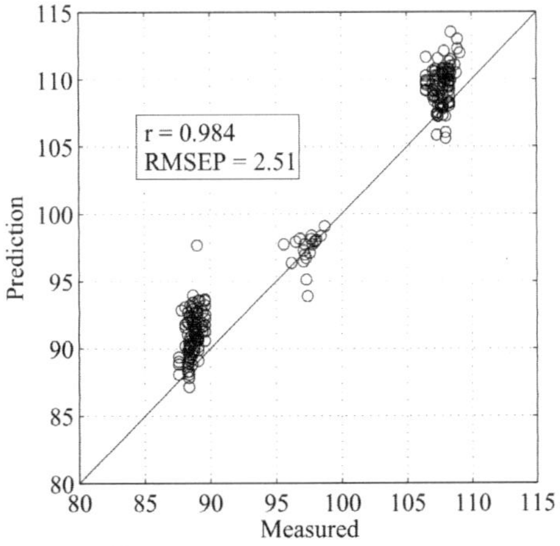

Figure 7: Estimation result (Case 7)

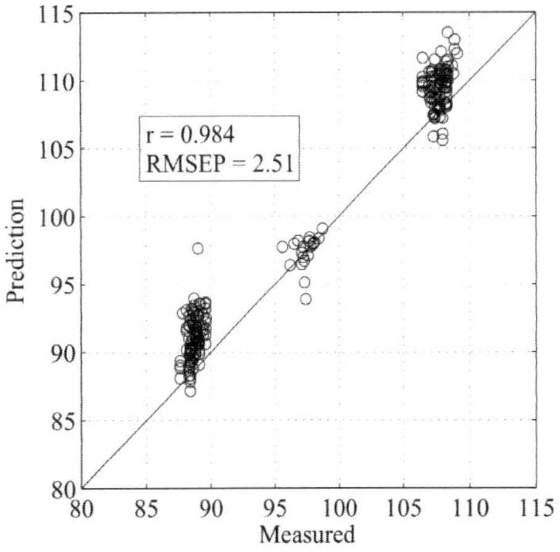

Figure 8: Estimation result (Case 8)

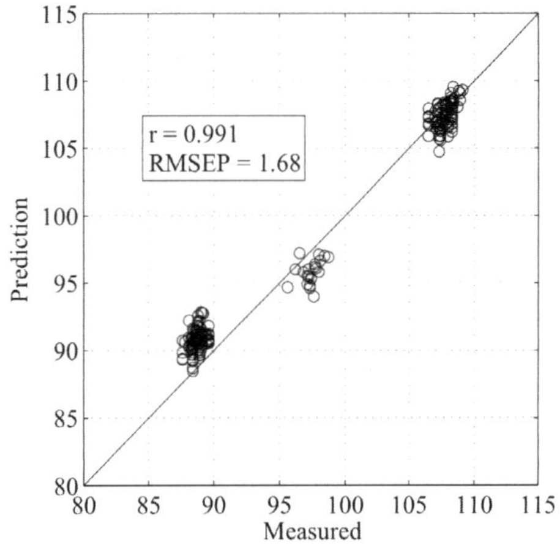

Figure 9: Estimation result (Case 9)

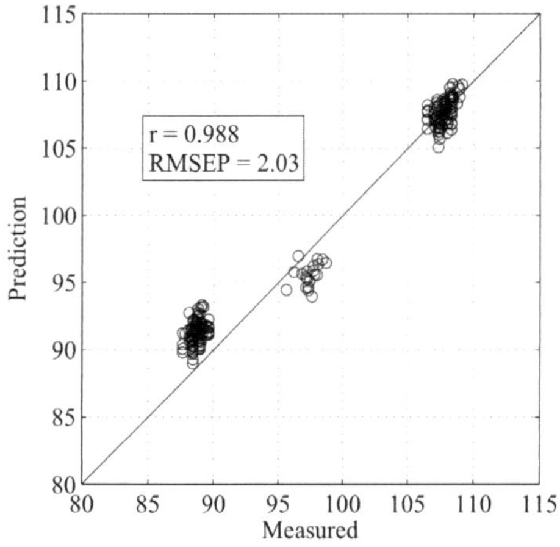

Figure 10: Estimation result (Case 10)

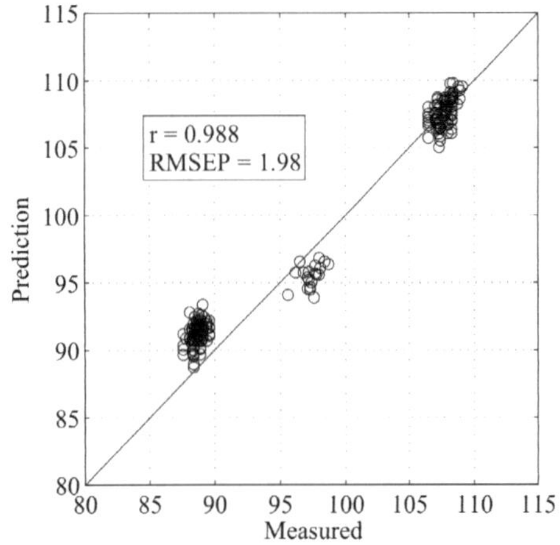

Figure 11: Estimation result (Case 11)

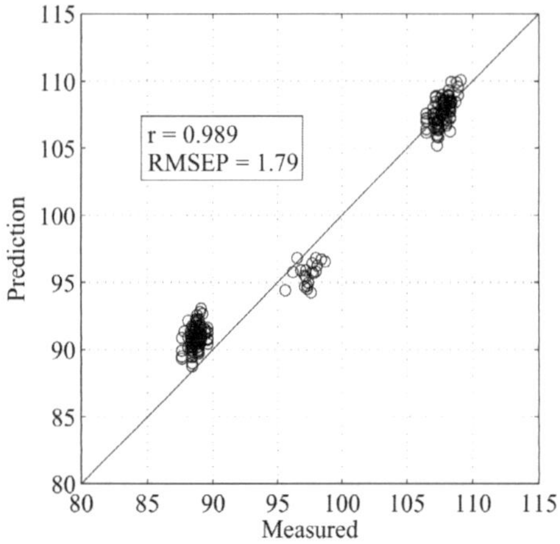

Figure 12: Estimation result (Case 12)

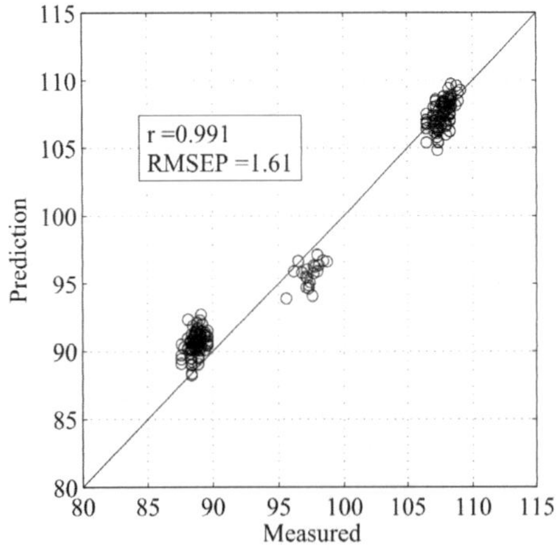

Figure 13: Estimation result (Case 13)

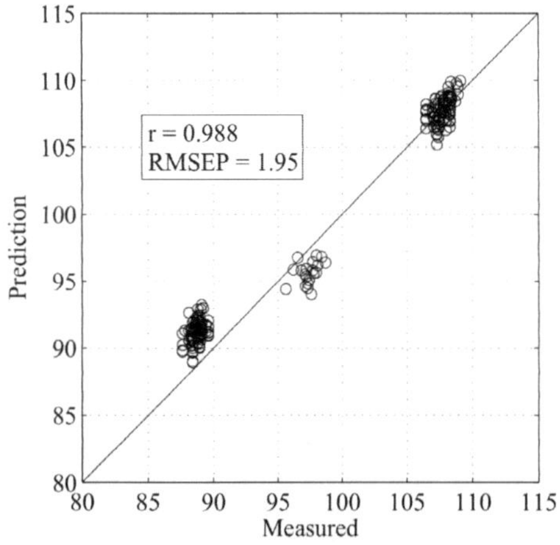

Figure 14: Estimation result (Case 14)

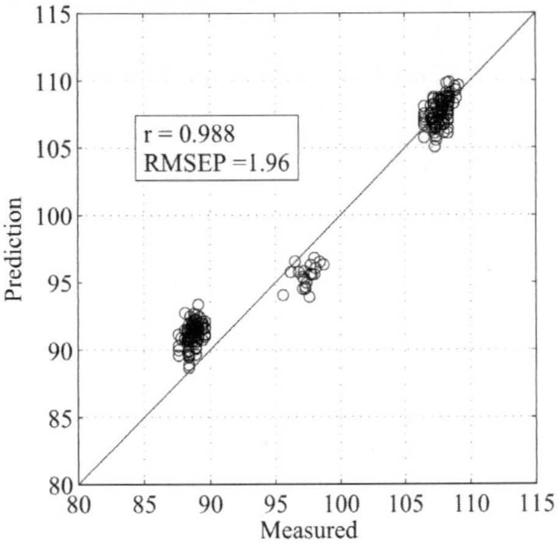

Figure 15: Estimation result (Case 15)

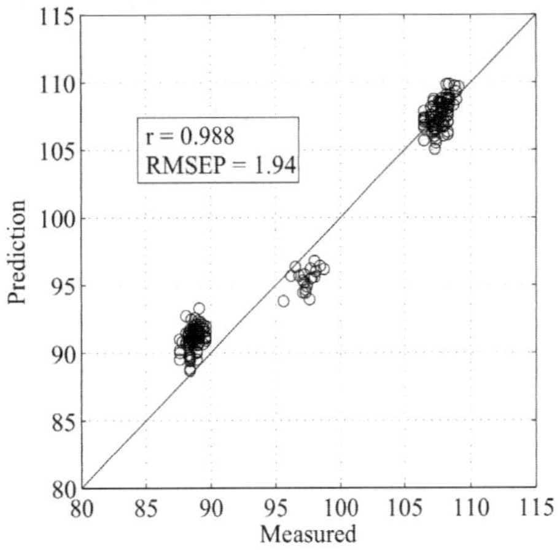

Figure 16: Estimation result (Case 16)

References

[1] Y. Roggo, P. Chalus, L. Maurer, C. Lema-Martinez, A. Edmond, N. Jent: A review of near infrared spectroscopy and chemometrics in pharmaceutical technologies, *Journal of Pharmaceutical and Biomedical Analysis*, **44**, 683/700 (2007)

[2] G. Reich: Near-infrared spectroscopy and imaging: Basic principles and pharmaceutical applications, *Advanced Drug Delivery Reviews*, **57**, 1109/1143 (2005)

[3] J. J. Moes, M. M. Ruijken, E. Gout, H. W. Frijlink, M. I. Ugwoke: Application of process analytical technology in tablet process development using NIR spectroscopy: Blend uniformity, content uniformity and coating thickness measurements, *International Journal of Pharmaceutics*, **357**, 108/118 (2008)

[4] H. Berthiaux, V. Mosorov, L. Tomczak, C. Gatumel, J. F. Demeyre: Principal component analysis for characterising homogeneity in powder mixing using image processing techniques, *Chemical Engineering and Processing*, **45**, 397/403 (2006)

[5] R. P. Cogdill, C. A. Anderson, M. Delgado-Lopez, D. Molseed, R. Chisholm, R. Bolton, T. Herkert, A. M. Afnan, J. K. Drennen III: Process analytical technology case study part I: Feasibility studies for quantitative near-infrared method development, *AAPS PharmSciTech*, **6(2)**, Article 37 (2005)

[6] H. Wu, M. Tawakkul, M. White, M. A. Khan: Quality-by-Design (QbD): An integrated multivariate approach for the component quantification in powder blends, *International Journal of Pharmaceutics*, **372**, 39/48 (2009)

[7] W. Li, G. D. Worosila: Quantitation of active pharmaceutical ingredients and excipients in powder blends using designed multivariate calibration models by near-infrared spectroscopy, *International Journal of Pharmaceutics*, **295**, 213/219 (2005)

[8] O. Berntsson, L. -G. Danielsson, B. Lagerholm, S. Folestad: Quantitative in-line monitoring of powder blending by near infrared reflection spectroscopy, *Powder Technology*, **123**, 185/193 (2002)

[9] Y. Sulub, B. Wabuyele, P. Gargiulo, J. Pazdan, J. Cheney, J. Berry, A. Gupta, R. Shah, H. Wu, M. Khan: Real-time on-line blend uniformity monitoring using near-infrared reflectance spectrometry: A noninvasive off-line calibration approach, *Journal of Pharmaceutical and Biomedical Analysis*, **49** ,48/54 (2009)

[10] S. Virtanen, O. Antikainen, J. Yliruusi: Uniformity of poorly miscible powders determined by near infrared spectroscopy, *International Journal of Pharmaceutics*, **345**, 108/115 (2007)

[11] A. U. Vanarase, M. Alcalá, J. I. Jerez Rozo, F. J. Muzzio, R. J. Romañach: Real-time monitoring of drug concentration in a continuous powder mixing process using NIR spectroscopy, *Chemical Engineering Science*, article in press

[12] D. Jouan-Rimbaud, D. L. Massart: Genetic algorithms as a tool for wavelength selection in multivariate calibration, *Analytical Chemistry*, **67**, 4295/4301 (1995)

[13] D. Jouan-Rimbaud, B. Walczak, D. L. Massart, I. R. Last, K. A. Prebble: Comparison of multivariate methods based on latent vectors and methods based on wavelength selection for the analysis of near-infrared spectroscopic data, *Analytica Chimica Acta*, **304**, 285/295 (1995)

[14] Z. Xiaobo, Z. Jiewen, M. J. W. Povey, M. Holmes, M. Hanpin: Variables selection methods in near-infrared spectroscopy, *Analytica Chimica Acta*, **667**, 14/32 (2010)

[15] R. J. Barnes, M. S. Dhanoa, and Susan J. Lister: Standard normal variate transformation and de-trending of Near-Infrared diffuse deflectance spectra, *Applied Spectroscopy*, **43**, 772/777 (1989)

(445e) Desensitizing near-Infrared Calibration Models for Physical Interferences in Tablets During Formulation/Process Development

Benoit Igne, Zhenqi Shi, Carl A. Anderson and James K. Drennen III, School of Pharmacy, Duquesne University, Pittsburgh, PA

During formulation/process developments, experimental designs are often used to minimize the number of experiments and maximize the necessary information on the chemical and physical stability of the drug, the flow properties, the process parameters, and the process monitoring tools, etc. As a consequence, the main effects of many parameters are often coupled with minor effects and it can become difficult to isolate phenomena of interest.

Using process analytical technologies (PAT) to monitor in real time blend homogeneity and tablet uniformity has become increasingly attractive in the pharmaceutical industry. Near infrared spectroscopy (NIRS) has predominantly been used since it can predict concentrations of drugs and excipients with reasonably low detection and quantification limits. NIRS can provide, during formulation/process development, valuable information to scientists regarding the chemical and physical properties of API and excipients, including powder flow behaviors, segregation issues, tablet properties (i.e. hardness), etc. However, the amount of variability present during formulation/process experimental designs (variations in excipient ratio, variation in blend time, variation in press parameters) can perturb NIRS calibration models and prevent the development of optimal models from monitoring precisely and accurately the process of interest. Variations in compression forces used to make tablets are one of the potential interferences that can limit the chemometrics model performances.

The present study proposes to desensitize near infrared calibration model to compression (physical property) information present in the calibration set to ease the interpretation of the results given by the design of experiment. Using partial least squares (PLS) and classical least square (CLS) regression models, various desensitization methods were implemented. Orthogonalization methods, aiming at removing the physical interferences were compared with the augmentation of the loading matrix of CLS and increasing the number of latent variables in the PLS model. Methods based on the net analyte signal were also developed to focus the calibration model on the parameter of interest and not include interferences. These methods were implemented along with an optimization of the spectral pretreatments.

Desensitizing PAT models to interfering phenomena can help process understanding and process control by focusing efforts on the parameters of interest and not implementing strategies to limit their effects. In the long run, these approaches will simplify the development of PAT methods and help the pharmaceutical industry move toward a Quality by Design framework. This study provides practitioners with a set of tools to ease the implementation of PAT methods during formulation development where the number of samples available is limited and many variables can interfere with building a precise and accurate multivariate model.

See more of this Session: Applications of Multivariate Data Analysis in Pharmaceutical Process Development
See more of this Group/Topical: Topical i : Comprehensive Quality by Design in Pharmaceutical Development and Manufacture

IMAGING PHARMACEUTICAL FLUIDIZED BED PROCESSING BY ELECTRICAL CAPACITANCE TOMOGRAPHY

Haigang Wang[1, 2], *Lubomir Gradinarsky*[3], *Livia Cerullo*[4], *Staffan Folestad*[3], *Wuqiang Yang*[1]

[1.] School of Electrical and Electronic Engineering, The University of Manchester, UK
wanghaig@hotmail.com, w.yang@manchester.ac.uk
[2.] Institute of Thermophysics Engineering, Chinese Academy of Sciences, Beijing, China
[3.] Pharmaceutical Development, AstraZeneca R&D Mölndal, Sweden
[4.] Chalmers University of Technology, Gothenburg, Sweden

Abstract

An electrical capacitance tomography (ECT) sensor with 8 electrodes was used to measure fluidized bed processes with different operating conditions. Key process parameters are analyzed, including process moisture, cross-sectional solids concentration, time-averaged raw capacitance, averaged solids concentration in the chamber, the effect of a Wurster tube on hydrodynamic behavior and the frequency spectra. The test results show that the change in the process moisture can be measured from the raw capacitance of adjacent electrode pairs combined with a calibration model. It is also shown that ECT could provide information on the fluidized bed process state and simultaneously could be used as a process monitoring and fault diagnostics tool. Future work is discussed in the end of the paper.

Keywords: Electrical capacitance tomography, Pharmaceutical fluidized bed, Drying, Coating

Introduction

Fluidized bed processes for drying, granulation and coating are important operation steps in the pharmaceutical industry. The quality of the finalized pharmaceutical product is strongly dependent on the proper setting of the operational parameters of these processes [1]. Due to the complex dynamics of the three-phase (gas-solids-liquid) fluid bed process, it is difficult to measure its behavior or diagnose faults using the conventional measurement tools, such as pressure or optical fiber sensors. For a multi-phase processes, non-intrusive imaging such as process tomography, can be a suitable alternative. Regarding the generally non-conductive nature of the pharmaceutical materials, ECT can be suitably applied in fluidized bed drying, granulation and coating processes. Some of the advantages of ECT are that it is fast (>100 frames per second) and can reconstruct permittivity distribution, which on the other hand is proportional to the location of the processed material in the fluid bed, as well as to the material density and moisture content. We studied the effect of presence and absence of a Wurster tube [3] inside the bed; the possibilities ECT to provide online monitoring. To achieve such objectives, an ECT sensor with 8 electrodes was used to monitor the fluidized bed processes with different operational conditions, where a subset was designed according to the Design of Experiments (DoE) principle. The utilized DoE factors were the inlet air flow rate and its temperature. The key investigated process characteristics included cross-sectional area solids concentration, time-averaged raw capacitance, averaged solids concentration in the chamber, the effect of the Wurster tube on hydrodynamic behavior and the frequency spectra derived from the ECT images.

The ECT sensor and system

The ECT system consists of a control box with 16 measurement channels [2]. The excitation frequency can be programmed in the range of 10 to 500 kHz and the voltage amplitude up to 20 $V_{p\text{-}p}$. Using a linear back-projection (LBP) algorithm, images can be reconstructed. When high-quality images are required, the Landweber iteration algorithm can be used for image reconstruction [4]:

$$G^{n+1} = G^{n} + \alpha \cdot (C_N - S \cdot G^{n}) \qquad (1)$$

where α is the step length or gain factor, G is the permittivity distribution and S is the sensitivity map.

The ECT sensor consists of 8 electrodes, which are made of adhesive copper tape and stuck onto vinyl tiles and to the fluidized bed chamber using double-sided foam tape as shown in Fig. 1 (a). The core of a coaxial cable was soldered to each electrode. The shield of the cable was connected to the stainless wall of earthed chamber. Fig. 1 (b) shows the set up in a lab at The University of Manchester with Sherwood M501 fluidized bed.

(a) ECT sensor (b) ECT system and fluidized bed

Fig.1 The ECT system and the test facility at The University of Manchester

Experimental results and analysis

Hydrodynamic test with two types of air distributor

Fig. 2 shows reconstructed images of solids concentration for two types of air distributor plates with different fluidization flow rate. The images clearly show the effect of the air distributor on the flow behavior in the fluidized bed. With a normal air distributor, air bubbles randomly move in different position in the chamber as shown in Fig. 2 (a). When the air distributor was half-blocked, most air bubbles move in one side of the chamber as shown in Fig. 2 (b). The test results show that the primary factor influencing the fluidized bed processing is the air flow and the air distributor [5] [6]. The test results also indicate that ECT can provide useful information for online monitoring and fault diagnosis, for example over-wetted particles stuck onto the air distributor [1].

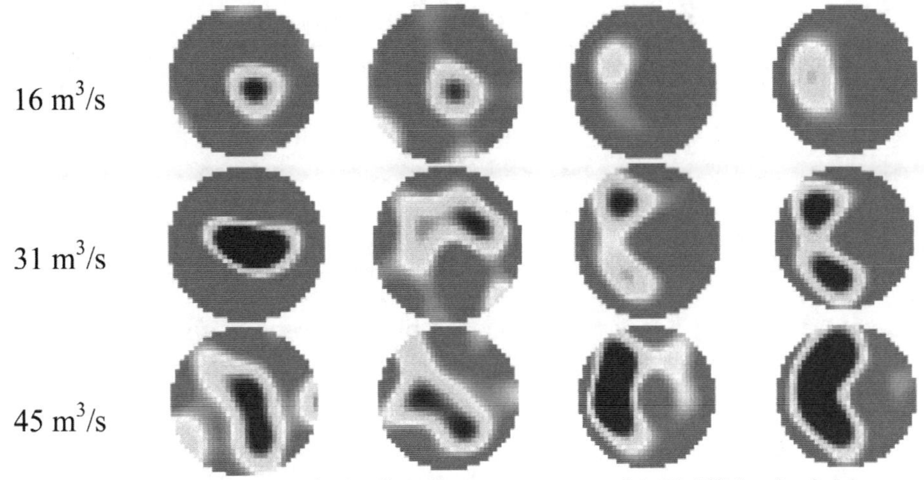

(a) Normal air distributor (b) Half blocked Air distributor

Fig. 2 Reconstructed images with different air distributor and different flow rate

Drying process

Fig. 3 shows the averaged raw capacitance for all of the electrical pairs and the change in the process moisture during a drying process with different process conditions selected in a DoE manner, where 5 types of precesses were defined (see table 1). The flow rate ranged from 16 to 45 m^3/h and the inlet temperature ranged from 30 to 80°C. From these curves, the effect of flow rate and temperature on the drying process can be seen. The drying time changed dramatically from 24 minutes as shown in Fig. 3 (f) to 100 minutes as shown in Fig. 3 (d) depending on the process parameters. Cases (b) and (c) are repeats.

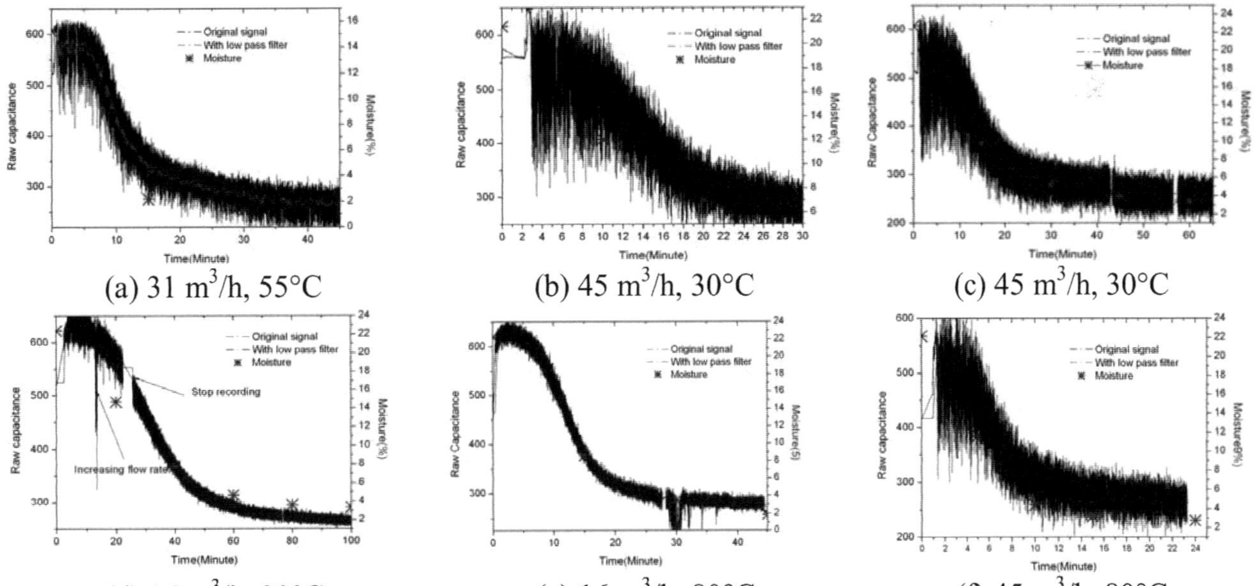

(a) 31 m^3/h, 55°C (b) 45 m^3/h, 30°C (c) 45 m^3/h, 30°C

(d) 16 m^3/h, 30°C (e) 16 m^3/h, 80°C (f) 45 m^3/h, 80°C

Fig. 3 Raw capacitance and the change in the moisture in a drying process (measured at-line by a loss on drying method) with different operational conditions selected in a DoE manner (air flow rate and its temperature as factors)

Table 1. The process conditions of the drying process

Case	Flow rate (m^3/h)	Inlet temperature (°C)
1	31	55
2	16	30
3	16	80
4	45	80
5	31	55

With the application of a low-pass filter, the process trend can be seen more clearly compared with the original signals as shown in Fig. 3. The procedure of the low-pass filter can be written as following,

$$y_i = \chi \cdot x_i + (1-\chi) \cdot y_{i-1} \qquad (2)$$

where $\chi = 1/(1+f/\Delta T)$ is a delay factor, f is the system frequency and △T is the delay time.

For all of the above cases, the signals decreased quickly in the first few minutes and then tended to decrease slower in the end of dying. The measured results agreed well with the drying theory indicating there are two drying stages for fine particles: a falling drying stage and followed by constant drying stage. [7].

Comparing the curves in Fig. 3 (d) and (e) with other curves, it can be seen that the fluctuation of the original signal with a low flow rate is smaller that that with a high flow rate. This can be seen more clearly in terms of spatial frequency spectrum as shown in Fig. 4. With a low flow rate, the dominant frequency is around 4.5 Hz. With a middle flow rate, it is around 2.8 Hz. With a high flow rate, there is no obvious dominant frequency. The above signals are taken in the beginning of drying with the same moisture level for the three cases.

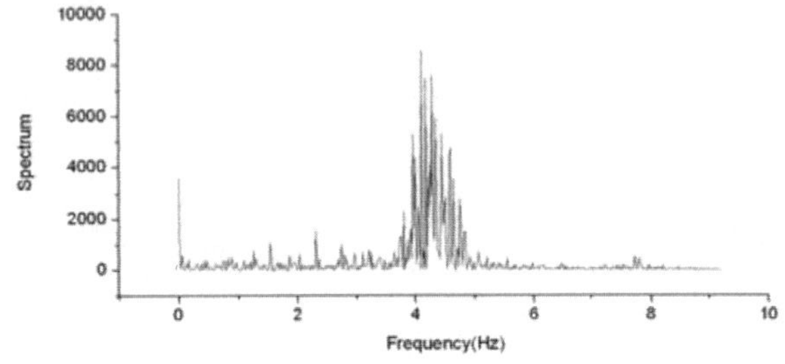

(a) 16 m^3/h

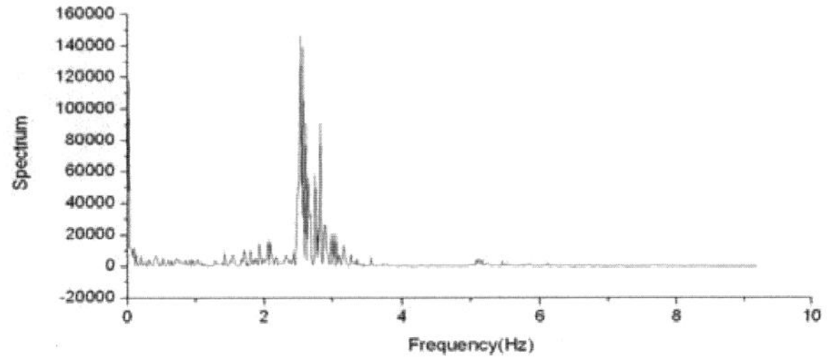

(b) 31 m^3/h

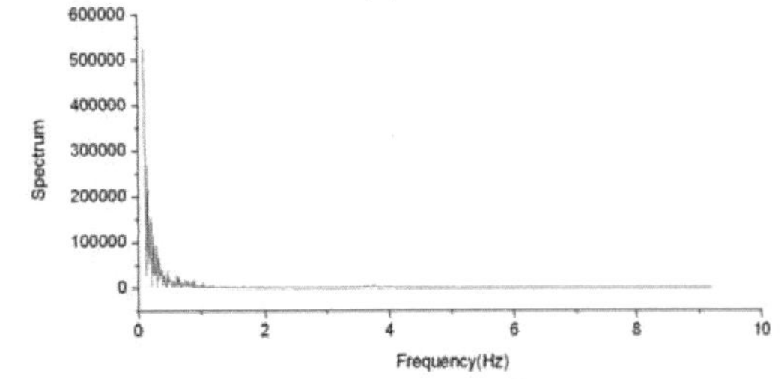

(c) 45 m^3/h

Fig. 4 Spatial frequency spectrum of the imaged capacitance for different flow rates

Fig. 5 shows the reconstructed images and the frequency spectrum for different drying times and for a constant low flow rate. The images show the change of the permittivity distribution. The dominant frequency changes from 4.5 Hz in the beginning to 8.5 Hz in the end of the drying process. It means that the flow behavior changes from slow fluctuation to fast fluctuation due to the change in the particle density with the drying time.

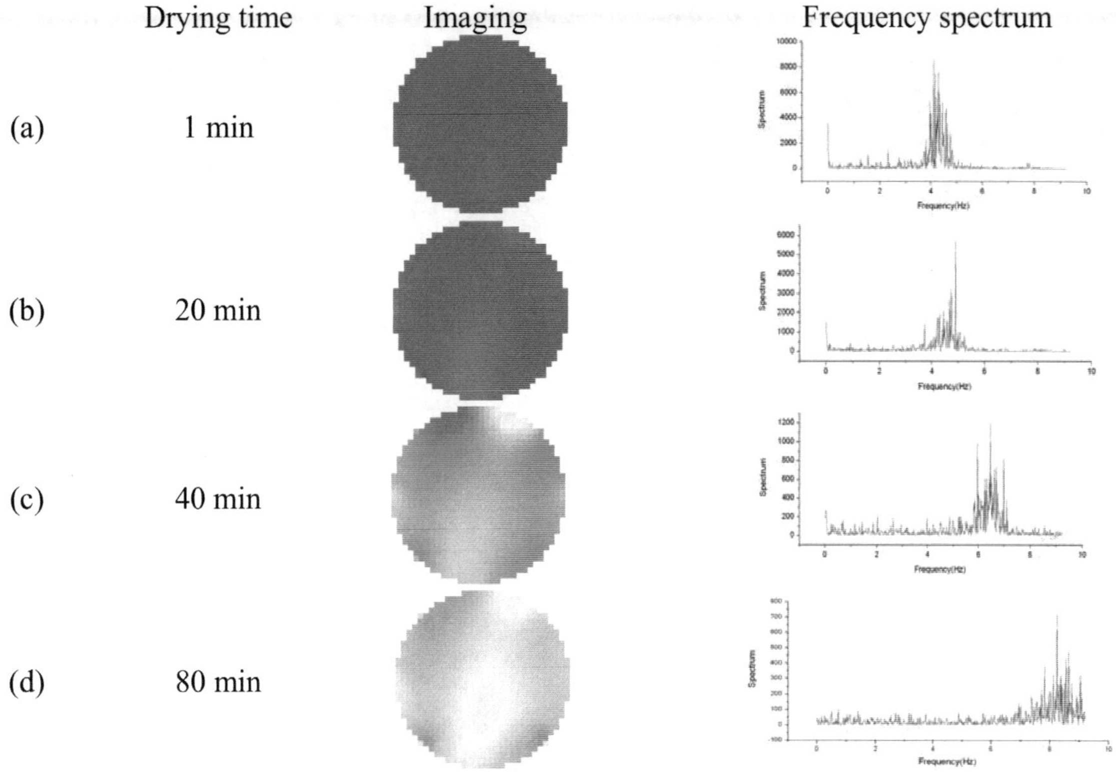

Fig. 5 Reconstructed images and frequency spectrum in different drying time (Flow rate: 16 m^3/h, inlet temperature: 30°C)

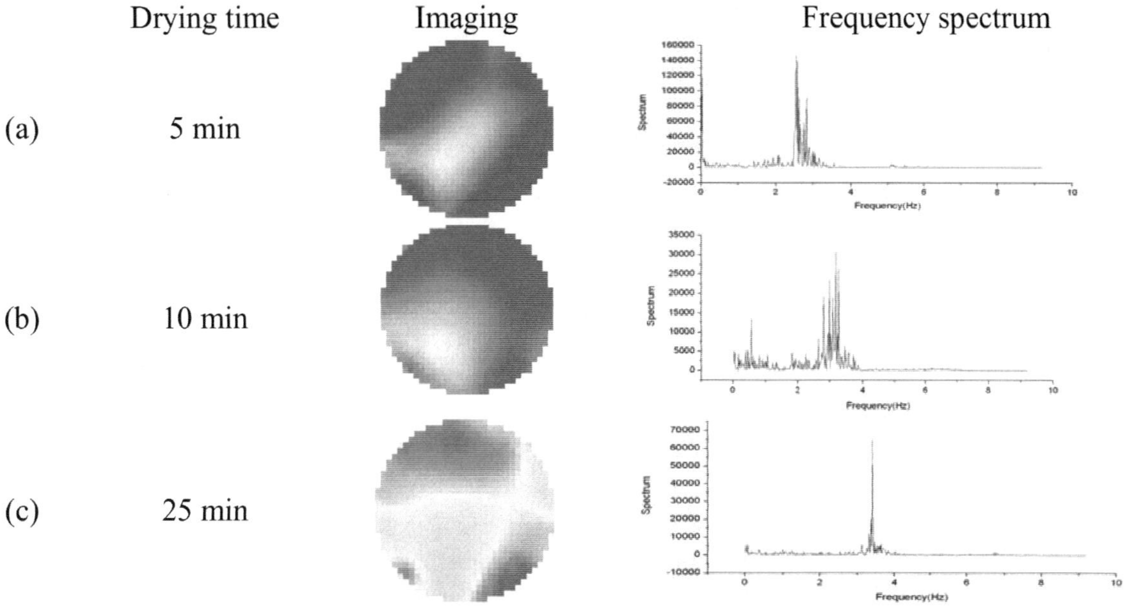

Fig. 6 Reconstructed images and frequency spectrum in different drying time (Flow rate: 31 m^3/h, inlet temperature: 55°C)

Fig. 6 shows the reconstructed images and the frequency spectrum in different drying time with a high flow rate. The frequency changes from 2.8 Hz in the beginning to 3.9 Hz in the end of the drying process. Comparing Fig. 6 with Fig. 5, the change in frequency is smaller with a high flow rate than that with a low flow rate.

Fig. 7 shows the drying rate calculated from the raw capacitance shown in Fig. 3. Two different drying conditions are shown In the first drying stage, a high flow rate results in a higher drying rate [8], because the mass and transfer coefficients with a high flow rate are bigger than that with a low flow rate. For the second process stage with the falling drying rate a high inlet temperature (55°C) causes higher drying rate compared to a low inlet temperature (30°C). This indicates that in the first drying stagees, a higher flow rate can be used while in the second stage a high drying temperature can be set.

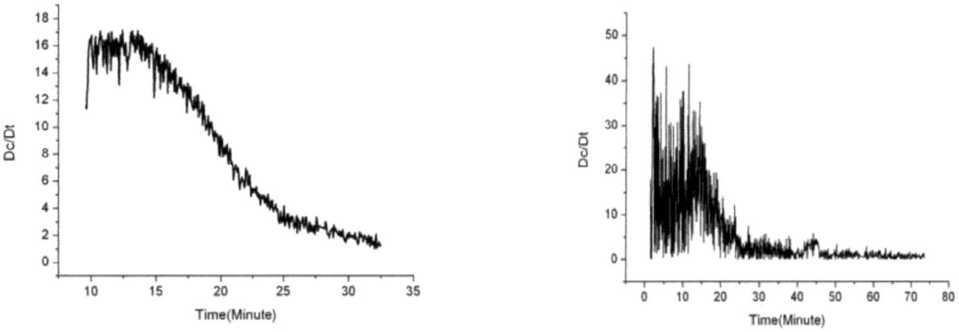

(a) 31 m^3/h, 55°C (b) 45 m^3/h, 30°C

Fig. 7 Drying rate with different flow rate and inlet temperature

Fig. 8 shows the averaged solids concentration along the radial direction with different fluidization velocities. Fig. 8 (a) shows the value in the beginning of the drying and (b) in the end of the drying. Due to a high moisture content in the beginning, the particles are difficult to fluidize with a low flow rate as shown in Fig. 8 (a). However, in the end of dying, due to the low moisture content, the density of the particles reduceds and it becomes possible to fluidize with a low flow rate.

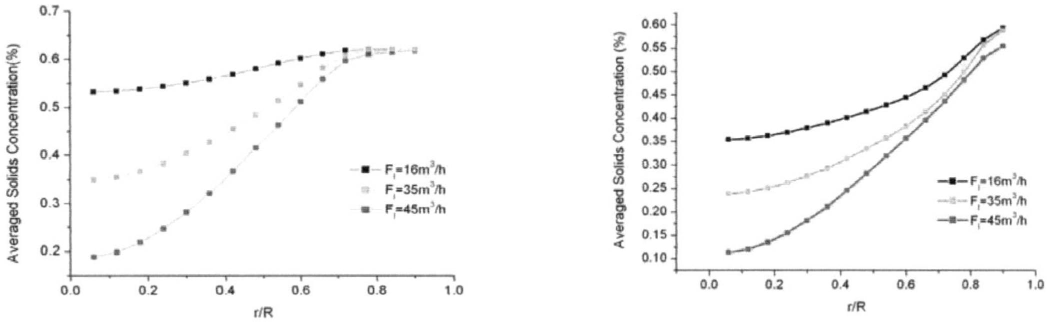

(a) In the beginning of drying (b) In the end of drying

Fig. 8 Averaged solids concentration with different fluidization velocity (particle weight: 600 g)

Fig. 9 compares two batches of drying with different weight of particles. For a high loading process as shown in Fig. 9 (a), the solids concentration presents a different profile for the different drying times. However, the solids concentration has the same profile both in the central region and near wall area for a low loading of pellets in different drying times. In terms of heat and mass transfer, there is an optimum solid's distribution and total weight for a specific drying chamber. Optimization of fluidised bed drying design and operation is planned as a future work.

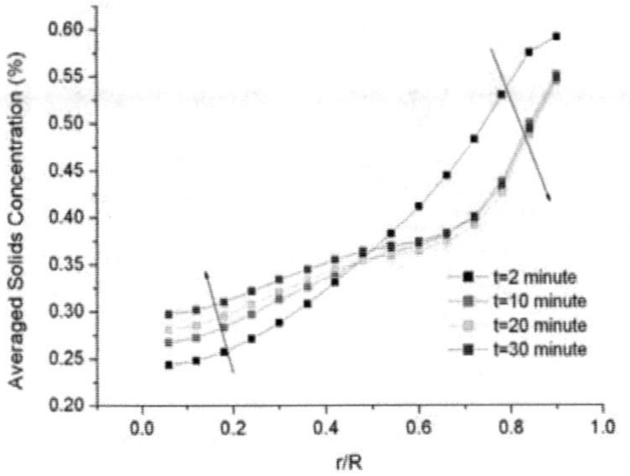

(a) 1 kg particles

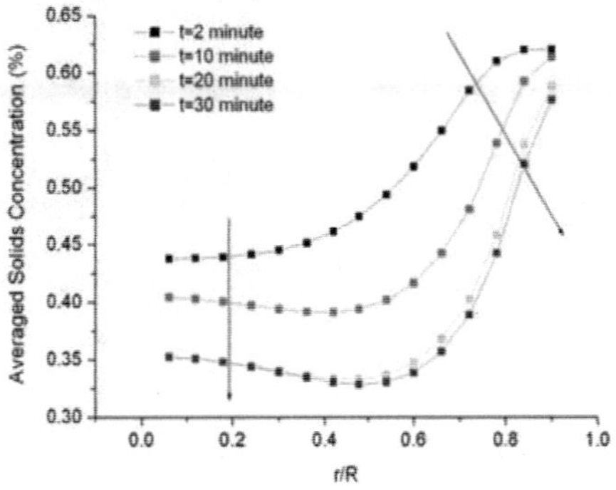

(b) 600 gram particles

Fig. 9 Averaged solids concentration at different drying times (r=0 is in the center of the bed, R is the bed radius)

Fluidized bed dryer with Wurster tube

In the following tests, two different lengths of Wurster tube as well as different atomizer air conditions were utilized. The lengths of the Wurster tubes used were 8.6 cm and 4.5 cm respectively. The automizer air pressure varied from 0 to 2.43 bar.

Fig. 10 and 11 gives the reconstructed images of solids distribution for the fluidized bed with a Wurster tube and under different operational conditions. When using a long Wurster tube, more air bubbles could be seen in the chamber in comparrioson to when the short tube is used. From the images, it is difficult to see the effect of the atomizer air pressure used on the solids distribution in the chamber in the two cases. The effect of the flow rate is also not obvious.

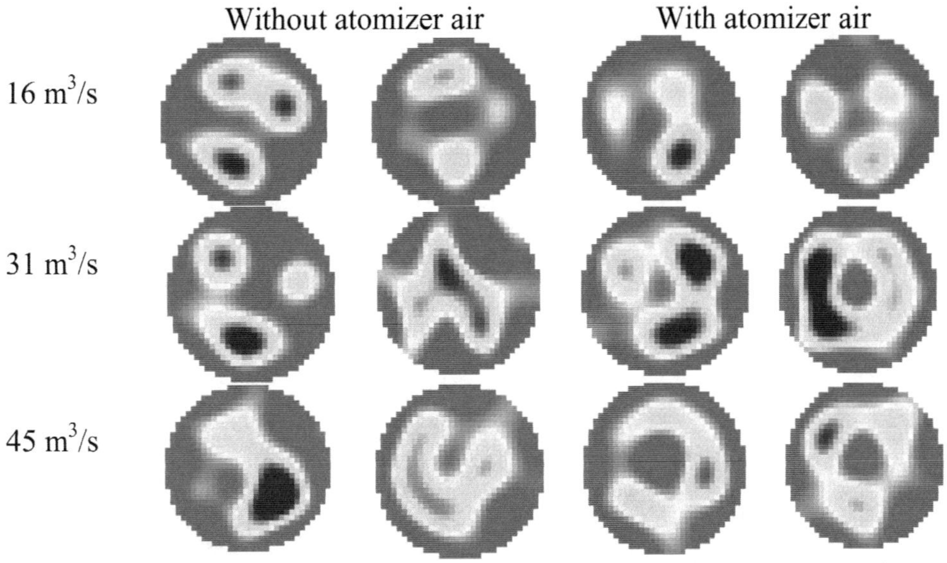

Fig. 10 Image reconstruction for fluidized bed with long Wurster tube

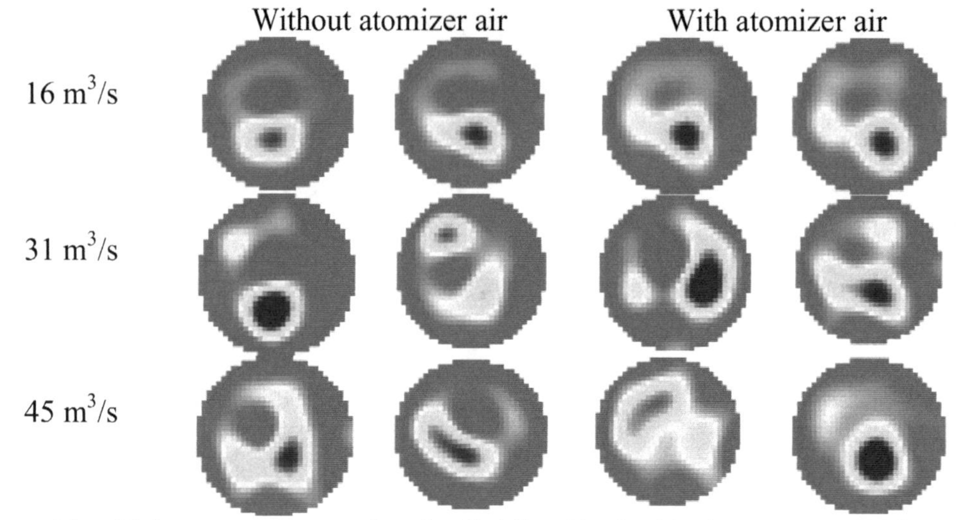

Fig. 11 Image reconstruction for fluidized bed with short Wurster tube

Fig. 12 shows the process curves for Wurster fluidized bed with different operation conditions. With a long Wurster tube, the effect of the atomizer air pressure on the average raw capacitance is clear, i.e., the data decrease with the increasing of atomizer air pressure (see Fig 12(a)). For a short Wurster tube, the effect is not as clear when compared to the case with usage of a long Wurster tube and therefore not displayed in Fig 12b. For both cases, the raw capacitance first increases with the increasing of flow rate (from 16 m^3/h to 31 m^3/h) and then decreases with the further increasing of flow rate (from 31 m^3/h to 45 m^3/h).

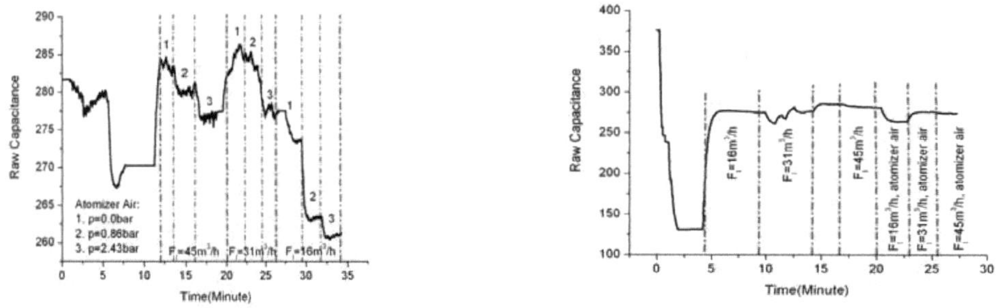

(a) Long Wurster tube (b) Short Wurster tube

Fig. 12 Capacitance reading with different Wurster tubes and atomizer air pressures

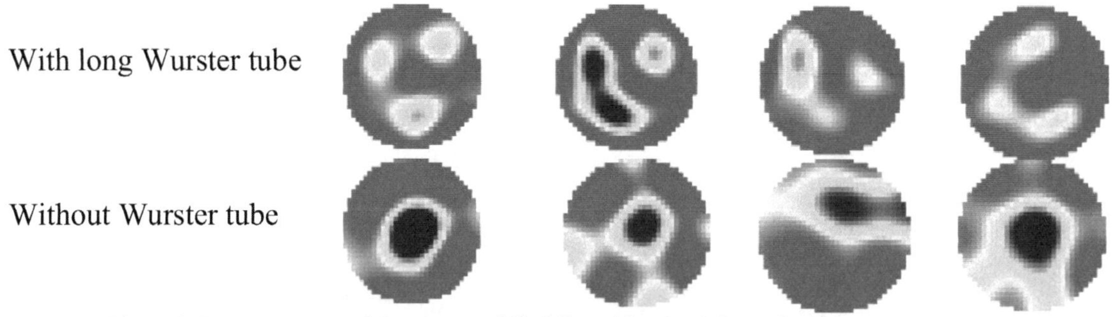

Fig. 13 Reconstructed images of fluidized bed with and without Wurster tube

Fig. 13 compares the reconstructed images with and without a Wurster tube and with the same process conditions. It is clear that the Wurster tube has effect on the flow behavior in the fluidized bed chamber and this can be detected by the ECT signals. However, further validation of the ECT results is needed where a more complex ECT sensor utilizing both the inside and the outside of the Wurster tube is planned to be tested. This will increase the sensitivity in the middle of the fluid bed.

Fig. 14 shows the change in the raw capacitance of a process with a Wurster tube where a few different phases were utilized. In the first phase a spraying process was simulated and we observed an expected increase in capacitance (time period of 2 to 15 minutes). The capacitance then deceases in the drying phase (after the 15th minute) returning to its original value at the end of the drying at the 46th minute. In the time period 46-52 min., the process was wet to an extreme condition, i.e. over spraying. In this stage, more and more water was sprayed into the fluidized bed and the particles stuck together until they stopped fluidizing. A process of recovery of the fluidization was observed after the spaying was stopped.

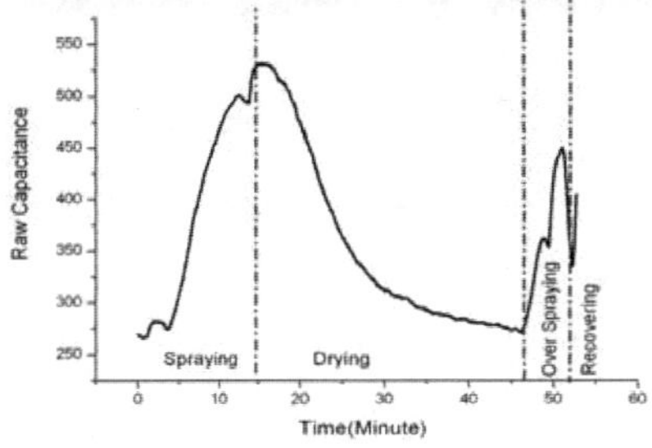

Fig. 14 Capacitance values for different spraying and drying processes for a bed with a Wurster tube

Fig. 15 shows the reconstructed images in time sequence. As it can be seen, the solids concentration increases first and then decreases until it recovers to the initial point in the end of drying. Due to more water spraying into the chamber, the particles become over-wet and the capacitance reach to the upper detection limit of the ECT system. The particles start to fluidize again after 54 minute when no further spraying of water occured.

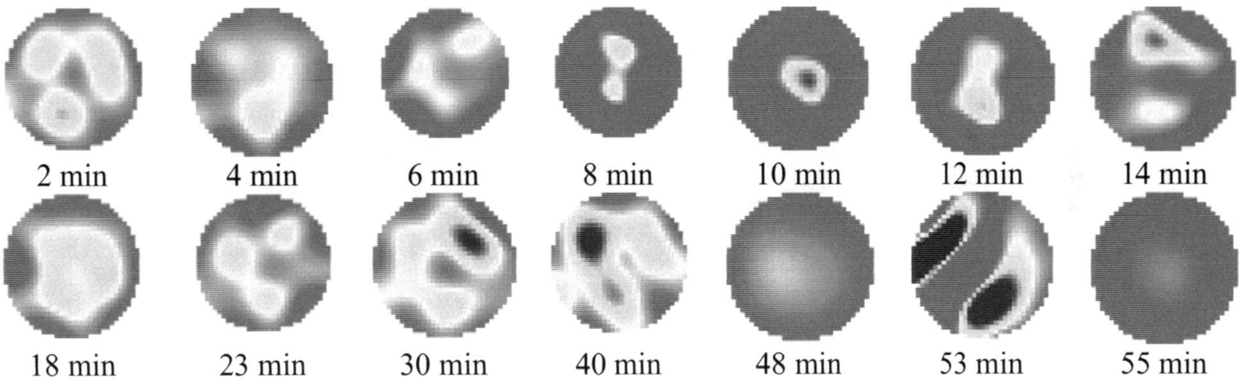

Fig. 15 Reconstructed capacitance images of the process from figure 16.

Measurement of the process moisture and evaluation of the consumed energy

Fig. 16 (a) shows the change in the process moisture for different operational conditions and determined by at-line LOD. The end point of the moisture content varied in the range of 1.87 % to 3.36 %, depending on the flow rate and drying temperature. The drying time dramatically changed from 24 minute to 100 minutes. Fig. 16 (b) gives the averaged capacitance from adjacent electrode pairs for the different drying conditions. At the process end point the measurements varied from 252.5 to 264.4. One notices that the absolute value of the capacitance changes depending on the process conditions but the curves in Fig 16 (b) present the same tendency as shown in Fig. 16 (a).

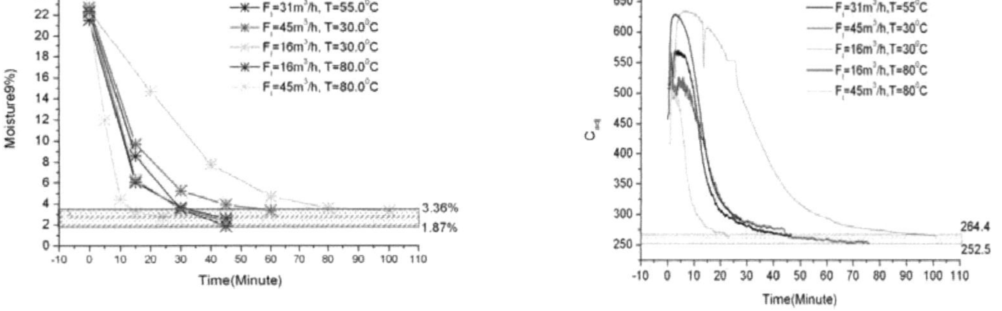

(a) Moisture (b) Adjacent electrode pair capacitance

Fig. 16 Moisture and capacitance vs. time for the different process conditions used in the DoE

Fig. 17 shows on the other hand the average capacitance and its gradient changes. Fig. 17 (a) shows a drying process with a middle flow rate and middle inlet temperature and (b) with a high flow rate and high inlet temperature. The fluctuation is strong in the drying beginning and tends to become smoother at the end of drying in both cases. With a high flow rate and high inlet temperature, the fluctuation is stronger than that with a middle flow rate and middle inlet temperature. We demonstrate that it is possible to use the gradient changes in order to determine the end point of drying.

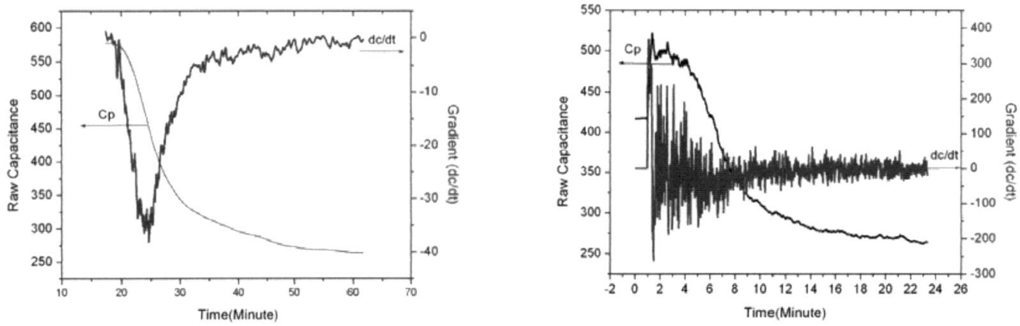

(a) 31 m^3/h, inlet temperature: 55°C (b) 45 m^3/h, inlet temperature: 80°C

Fig. 17 Capacitance and its gradient change

Conclusions

This paper presents the preliminary test results with ECT for monitoring of drying and spraying processes in fluidized beds and conduced at AstraZeneca in Mölndal, Sweden. The following can be concluded.

- The results demonstrated that ECT can be used with pharmaceutical fluidized beds for monitoring of drying processes. Further work is needed to optimize the electrode size and position, as well as the frequency and the amplitude of the excitation signal.

- The reconstructed images of the capacitance show that ECT can be used as an in-line tool for measureing the change in the solids' moisture. Raw capacitance signals can provide process information, regarding e.g. the nozzle working conditions and the process moisture. Further work is needed to find the relationship between the particle moisture, the solids concentration and the raw capacitance.
- Frequency spectrum analysis was useful to identify process flow regimes and monitor process conditions in a fluidized bed dryer.
- The time differentiation of original signals could be utilized to indicate the end point of the drying.
- Re-engineering of the ECT sensor is the key to the commercial success as applied to industrial fluidized beds. Since adhesive was used to stick the electrodes to the bowl wall, many adhesives loose adhesion at higher temperature.

Acknowledgments

The authors would like to thank EPSRC (GR/T29383/01, GR/T29376/01, EP/G005702/1) and University of Manchester Intellectual Property (UMIP) Ltd for financially supporting this research. AstraZeneca in Mölndal in Sweden is thanked for providing the fluidized bed and supporting this research. GEA Pharma System Ltd in Eastleigh (UK) and Sherwood Scientific Ltd in Cambridge (UK) are thanked for their collaboration.

References

1. Wang H.G., Yang W.Q., Proctor I., Taylor J., Marr A. and Page T.G. (2009), Online monitoring and flow regime identification of fluidised bed drying and granulation processes, IEEE International Workshop on Imaging Systems and Techniques, May 11-12, Shenzhen, China, pp 247-251
2. Yang W.Q. and York T.A. (1999), New AC-based capacitance tomography system, IEE. Proc.-Sci. Meas. Technol., 146, pp 47-53
3. Teunou E. and Poncelet D. (2002), Batch and continuous fluid bed coating- review and state of the art, J. of Food Engineering, 2002, 53, pp 325-340
4. Yang W.Q. and Peng L.H. (2003), Image reconstruction algorithms for electrical capacitance tomography, Meas. Sci. and Technol., 14, pp R1-R13
5. Depypere F., Pieters J. G. and Dewettinck K. (2004), CFD analysis of air distribution in fluidised bed equipment, Powder Technol., 145, pp 176-189
6. Qureshi A.E. and Creasy D.E. (1979), Fluidised bed gas distributors, Powder Technol., 22, pp 113-1199
7. Vojtěch V.C., Markvart M. and Drbohlav R. (1966), Fluidised bed drying, Leonard Hill, London
8. Kunii D, and Levenspiel O (1991), Fluidization engineering, Butterworth-Heinemann, Boston
9. Yang W.Q. (2010), Design of electrical capacitance tomography sensors, Meas. Sci. and Technol., 21, 042001
10. Wang H.G., Senior P., Mann R. and Yang W.Q. (2009), Online solids moisture measurement and optimum control of fluidised bed dryer, Chem. Eng. Sci., 64, pp 2893-2902

(510a) Quality-by-Design (QbD): Integration of Design of Experiments (DOE) and Real-Time PAT Monitoring for Pharmaceutical Powder Blending Process Evaluation

Huiquan Wu, Division of Product Quality Research, HFD-940, FDA, Silver Spring, MD and Mansoor A. Khan, Cder/Ops/Otr/Dpqr, Food and Drug Administration, Silver Spring, MD

Although pharmaceutical powder blending process as an important unit operation has received significant attention, our understanding to some of the critical issues such as blending homogeneity assessment and blending kinetics evaluation is limited. As highlighted in the FDA's PAT Guidance, significant opportunities exist for real-time process monitoring and control using PAT tools such as Design of Experiments (DOE), Process Analyzers, and Chemometrics. In this work, efforts were made to explore the impact of both formulation variables and process variables on powder blending performance measurement in a high-shear mixer, using Near-Infrared (NIR) real-time process monitoring and design of experiments (DOE) approach. Ibuprofen was used as a model drug. MCC was used as a model excipient. First, a full factorial design was created to examine the effects of blending process variables (high-shear impellor speed and probe insertion depth) on the NIR spectra of pure component MCC in the KG-5 high-shear mixer. Second, a full factorial design was created to examine the effects of both formulation variables (ratio of Ibuprofen and MCC) and process variables (impellor speed and probe insertion depth) on powder blending performance measured via NIR real-time process monitoring. The first DOE study demonstrated the measurement robustness of the real-time NIR process monitoring and provided evidence of the experimental errors expected in this kind of study protocol. The second DOE study demonstrated how formulation variable and process variables impact the powder blending performance evaluation via NIR real-time process monitoring. The knowledge gained through this study could provide scientific justifications for necessary powder blending process control and optimization in high-shear mixer.

See more of this Session: Applications of PAT in a Manufacturing Setting
See more of this Group/Topical: Topical i : Comprehensive Quality by Design in Pharmaceutical Development and Manufacture

(510b) Raman Spectroscopy Combined with Small and Wide Angle X-Ray Scattering as a Non Destructive Quality Control Tool for Powder Compression Analytics in Pharmaceutical Applications

Nicolas Heigl, Research Center Pharmaceutical Engineering GmbH, Graz, Austria and Johannes G. Khinast, Research Center Pharmaceutical Engineering (RCPE), Graz, Austria

This study evaluates the feasibility of a combined use of small-wide angle X-ray scattering (SWAXS) and Raman spectroscopy as a non-invasive process analytical tool (PAT) for the analysis of pharmaceutical granulation and compression procedures. Placebo formulations containing Lactose monohydrate (LM, 49 wt %), microcrystalline cellulose (MCC, 49 wt %), silicone dioxide ($SiO2$, 1 wt %) and magnesium stearate (MgSt) were subjected to iterative treatment cycles comprising compression, sieving and reaming. The morphology of the raw powders, and characteristics of the respective compressed pellets, was examined after every treatment cycle. Correlation analysis among compression force (F) and tablet hardness (H) revealed an increasing linearity, indicated by the correlation coefficient (r), with continuative granulation from 0.95 to 0.98. The SAXS spectra showed incremental particle-particle interfaces by multiple granulations whereas the WAXS data indicated no changes in morphology during the treatment procedures. Generally, with increasing powder compaction force (i) the particles consolidated and (ii) the surface smoothened. The Raman findings were in good agreement with the SWAXS data, meaning that a decrease in Raman backscattering intensity with enhanced powder compaction force was observed. Multivariate data analysis (MVDA) modeling in form of principal component analysis (PCA) and partial least squares (PLS) regression was used to qualitatively study and quantitatively obtain information from the Raman spectra. The multivariate correlation analysis among the X-variables (Raman spectra) and Y-variables (tablet hardness) showed the same tendencies as the linear correlation analysis, i.e. the validation set (v-set) r constantly increased from 0.88 to 0.96. The final PLS model for determining the hardness of unknown samples showed a root mean squared error (RMSE) of calibration (RMSEC) of 10.12 and prediction (RMSEP) of 12.88, respectively. These findings indicate that by combining SWAXS and Raman spectroscopy into one approach a very powerful and useful new process analytical tool would be available for future pharmaceutical quality control implementations.

See more of this Session: Applications of PAT in a Manufacturing Setting
See more of this Group/Topical: Topical i : Comprehensive Quality by Design in Pharmaceutical Development and Manufacture

(510c) On-Line Monitoring of Roller Compaction Ribbon Density Using NIR Spectroscopy

Ryan McCann[1], Ariel Muliadi[2], Janne Paaso[3], James Litster[4] and Rodolfo Pinal[1], (1)Industrial and Physical Pharmacy, Purdue University, West Lafayette, IN, (2)Mechanical Engineering, Purdue University, West Lafayette, IN, (3)Optical Instruments, VTT Technical Research Centre of Findland, Oulu, Finland, (4)Chemical Engineering, Purdue University, West Lafayette, IN

Solid dosage forms, such as tablets and capsules, have been predominately manufactured through batch processes in pharmaceutical industry for years. The development of continuous manufacturing processes is of high interest to the pharmaceutical industry to reduce the huge costs associated with process scale-up. Roller compaction is a dry granulation technique which is readily applicable to continuous processing and is currently a part of a test bed at Purdue University to demonstrate the feasibility of manufacturing tablets continuously. PAT implementation is a key component to the success of the continuous process since we must now measure critical quality parameters on-line. Near infrared spectroscopy has been selected as the analytical "eye" for monitoring ribbon density since it is an analytical technique which is non-destructive, fast, and measures both chemical and physical information about the sample.

Previous research has shown that roller compacted ribbons have a three dimensional density distribution which can affect the particle size distribution of granules after milling. Our specific interest is in understanding and monitoring the density distribution of ribbons produced by roller compaction with near infrared spectroscopy. We will also investigate the distribution of active pharmaceutical ingredients within the ribbons. Therefore, by understanding the effects of formulation, operating conditions, and equipment design on ribbon density distribution we can develop optimized monitoring strategies to help produce a more uniform final product.

The active pharmaceutical ingredient, acetaminophen (APAP), and excipients, including microcrystalline cellulose (MCC), magnesium stearate (MgSt) and Cal-O-Sil® (SiO_2) were selected as the model system to study the roller compaction process. To understand the effect of formulation on ribbon density distribution acetaminophen concentration, microcrystalline cellulose grades, and lubricant levels were changed. The effect of operating conditions on ribbon density distribution was studied by changing the roll force and the feed screw speed to roll speed ratio. Finally, the effect of equipment design was studied by altering the penetration of the feed screw into the slip region of the rolls. A multi-point, non-contact diffuse reflectance near infrared probe was set-up at the outlet of the rolls to interrogate the entire width of the ribbon. Offline envelop density measurements were used to create and confirm a multivariate density calibration model for each formulation. The multi-point spectra were combined together to create density contour maps.

In this work, we show the results of the multi-point experiments to characterize ribbon density distribution and acetaminophen distribution. We will also discuss how the variance of ribbon density changes with the scale of scrutiny. Based on the analysis of density variation, a near infrared monitoring strategy will be outlined. Finally, we will offer some general conclusions about how the experimental results can be used and extended to process control.

See more of this Session: Applications of PAT in a Manufacturing Setting
See more of this Group/Topical: Topical i : Comprehensive Quality by Design in Pharmaceutical Development and Manufacture

(510d) Real-Time Monitoring of Pharmaceutical Manufacturing Processes with a Multiplexed Optical Fiber Probe NIR-Setup

Nikolaus Balak[1], Daniel M. Koller[1] and Johannes Khinast[2], (1)Process Engineering and Manufacturing Science, Research Center Pharmaceutical Engineering GmbH, Graz, Austria, (2)Institute for Process and Particle Engineering, Graz University of Technology, Graz, Austria

Diffuse reflectance near-infrared (NIR) spectroscopy is a non-invasive and widely used technology with the potential for a wide range of applications in the pharmaceutical industry, ranging from raw material identification to final product qualification. Especially powder blending processes, as one of the key unit operations for solid dosage forms can be monitored in real-time for the blend quality and blending time until uniformity. Common NIR systems enable single point measurements at dedicated positions of the blender, based on the assumption, that the sampled position represents the whole blend. This is a problematic task as the mixing characteristics of powders significantly change (e.g., segregation, agglomeration, etc.) with the process parameters. Here we introduce, a new process analytical technology (PAT) setup for a spatially resolved quasi-simultaneous process monitoring for batch and continuous manufacturing processes. Therefore, an FT-NIR spectrometer was combined with an optical multiplexer, simultaneously switching the excitation and detection path of the bifurcated fibers, which can be connected to a blender via sapphire sampling ports at different positions. Different multi-component powder mixing processes were investigated and quantitatively analyzed with multivariate data analysis (MVDA) using software packages of CAMO and Umetrics, which also support real-time process monitoring. The new flexible PAT setup presented here, enables non-invasive in-line and real-time monitoring of dynamic processes, especially where spatial deviations can be expected e.g., segregation in solids (i.e., powder blends) and liquids (i.e., sedimentation).

See more of this Session: Applications of PAT in a Manufacturing Setting
See more of this Group/Topical: Topical i : Comprehensive Quality by Design in Pharmaceutical Development and Manufacture

(510e) On-Line Blend Monitoring of Pharmaceutical Materials Using Multiple NIR Sensors

Benoit Igne, Zhenqi Shi, Carl A. Anderson and James K. Drennen III, School of Pharmacy, Duquesne University, Pittsburgh, PA

In recent years, powder blend monitoring and control systems using process analytical technologies (PAT) have become increasingly popular in the pharmaceutical industry. Among them, spectroscopic techniques allow for real-time evaluation of blend homogeneity and powder behaviors necessary for both formulation and process scientists to gain process understanding without invasive sampling to ultimately improve manufacturing quality. Near infrared spectroscopy (NIRS) has predominantly been used for blend homogeneity monitoring and control since it can predict in real time concentrations of drugs and excipients with reasonably low detection and quantification limits.

At small scales, the use of a single sensor might be sufficient to sample a representative fraction of a blend container, but at larger scales, it might be necessary to employ multiple sensors. Indeed, the scale of scrutiny of one sensor might not be large enough to provide an accurate estimate of the homogeneity of a blend in large scale manufacturing facilities. However, as additional sensors are added, an increased number of parameters must be optimized and a more complex decision strategy must be implemented. It is conceivable that one sensor might indicate the blend is homogeneous, while a second sensor might not have reached that same homogeneity state due to blender loading discrepancies, blender shape, multivariate model differences, spectrometer differences, etc.

The present study proposes to investigate the implementation of blend monitoring using two NIR sensors on a V-blender. In a first part, using a six-component formulation, the effect of prediction model error and spectrometer differences on blend end point determination will be studied. Discrepancies between sensor outputs will be evaluated and methods to limit their impact presented to ease the decision to stop blending. In a second part, the effect of loading patterns will be studied on a three-component formulation and their impact on the blend end point will be investigated.

The objective of this study is to provide practitioners with a set of tools to ease the implementation of multiple sensors during the blending process. PAT calibration model quality and its impact on the blend end point will be assessed as well as the need for calibration transfer or standardization methods to limit the effect of discrepancies between PAT sensors.

See more of this Session: Applications of PAT in a Manufacturing Setting
See more of this Group/Topical: Topical i : Comprehensive Quality by Design in Pharmaceutical

(510f) Drying Mechanism and Aggregate Formation of API

Daniel Hsieh, Joshua Engstrom, Deniz Erdemir, Steven Chan, Steve S. Y. Wang, Shih-Ying Chang, Chiajen Lai and San Kiang, Pr&d, Bristol-Myers Squibb Company, New Brunswick, NJ

After crystallization, the active pharmaceutical ingredient (API) isolated from a solvent/antisolvent system can form into hard aggregates during drying which could adversely affect drug product formulation. The objective of this study is to elucidate the mechanism of aggregate formation during drying of an API in the presence of a solvent/antisolvent system and to propose solutions to eliminate such formation. An integrated approach to understand and solve this problem was focused on 3 areas: crystallization, filtration/washing and drying. For the drying area, theoretical calculations were made to establish drying models that were verified experimentally with vapor concentration measurements made by mass spectroscopy and with drying curves formed by analyzing the wet cake weight change during drying. On the basis of these studies, pilot plant filtration/washing and drying protocols were modified which resulted in the elimination of hard aggregate formation. A mechanistic understanding of the formation of hard aggregates has wide applications to any API isolated from a solvent/antisolvent system.

See more of this Session: Applications of PAT in a Manufacturing Setting
See more of this Group/Topical: Topical i : Comprehensive Quality by Design in Pharmaceutical Development and Manufacture

(510g) Real-Time Monitoring of Continuous Processes with Spectroscopic PAT-Tools

Daniel M. Koller[1], Nikolaus Balak[1], Simon D. Fraser[1] and Johannes G. Khinast[2], (1)Process Engineering and Manufacturing Science, Research Center Pharmaceutical Engineering GmbH, Graz, Austria, (2)Institute for Process- and Particle Engineering, Graz University of Technology, Graz, Austria

The aim of Process Analytical Technology (PAT) is to gain deeper insight in pharmaceutical manufacturing processes, replacing empirical approaches by knowledge-based procedures. A detailed understanding of the key parameters of a process and their impact on the product quality allows companies to "build in quality" instead of "testing it into the final product". From this point of view, PAT has to be tailored for a direct implementation to manufacturing lines to measure in real-time the key parameters of the process.

Here, we present results on real-time quantitative in-line monitoring of powder mixing processes by near-infrared (NIR) spectroscopy [1, 2]. Spectral data collection and data analysis was performed in real-time based on partial least squares models developed via dynamic calibration. Different strategies are followed: I) Application of a single reflection probe to determine the mixing end-point. II) A novel multiple reflectance probe measurement system for a spatially resolved monitoring of the powder flow kinetics within a blender. The parameters from the mixing process are linked to keyparameters of powders, obtained from rheometric investigations based on experimental design. Beside the batch processes monitored here with NIR spectroscopy, we also introduce hyper spectral imaging for monitoring fast continuous processes and non-invasive product quality analysis of solid dosage forms.

[1] Koller D M, Posch A, Hörl G, Voura C, Radl S, Urbanetz N, Fraser S D, Tritthart W, Reiter F, Schlingmann M, Khinast J G. Continuous Quantitative Monitoring of Powder Mixing Dynamics by Near-infrared Spectroscopy. Powder Technology 2010; submitted

[2] Remy B, Glasser B J, Khinast J G. The effect of mixer properties and fill level on granular flow in a bladed mixer. AIChE Journal 2010, 56: 336-353. doi:10.1002/aic.11979

See more of this Session: Applications of PAT in a Manufacturing Setting
See more of this Group/Topical: Topical i : Comprehensive Quality by Design in Pharmaceutical Development and Manufacture

An Integrated Quality-by-Design (QbD) Approach towards Design Space Definition of Key Unit Operations In the Manufacturing of Solid Dosage Forms by Discrete Element Method (DEM) and Computational Fluid Dynamics (CFD) Simulation

Siegfried Adam[1], Daniele Suzzi[1], Stefan Radl[2], Gregor Toschkoff[1], Charles Radeke[1], Thomas Hoermann[1], Marco Iannuccelli[1] and Johannes Khinast[2], (1)Research Center Pharmaceutical Engineering (RCPE), Graz, Austria, (2)Institute for Process and Particle Engineering, Graz University of Technology, Graz, Austria

Combined Quality-by-Design and computer simulation approaches (discrete element method and computational fluid dynamics) are presented to characterize key unit operations of solid dosage form manufacturing, namely powder blending and tablet coating. The aim is to evaluate the impact of formulation parameters and process variables on process and product quality. Fundamental scientific knowledge is required to develop and optimize pharmaceutical manufacturing processes that yield constantly high quality products and provide most cost efficient manufacturing strategies. Understanding the variability of both the material attributes and process parameters, as well as their overall impact on the three key unit operations are critical elements for QbD. In a first step, the QbD-methodology is systematically used to (1) establish the critical quality attributes representative for the selected dosage forms, (2) identify potentially critical input factors that may affect process and product quality and (3) risk-rank these factors to define activities for process characterization. Subsequently, computer simulation-based characterizations of the three key unit operations are performed. Data are then used to map out three-dimensional knowledge spaces, providing parameters to define a design space as their subset and set up an appropriate control strategy.

Powder blending, being one of the key unit operations for solid dosage form manufacturing, strongly impacts the quality of the final product, and therefore, closely relates to safety and efficacy. In this study we develop in-depth understanding of how formulation variables of a model two-component formulation and the blending process parameters affect the blending kinetics, the blending end-point and the final blending quality. In our approach *blending homogeneity* was selected as surrogate for the critical quality attribute *content uniformity* of the final product, as it contributes to the overall product quality and process performance. By performing a risk-based approach, several causes for poor blending are identified. To demonstrate the validity of the simulation approach, parameters with partly-known effects on blending quality have been selected for further characterization. Thus, a QbD-based model is presented that relates blending time and concentration of the active pharmaceutical ingredient to blending quality [1]. Depending on these two variables, the blending end point and the blend quality can be predicted.

Tablet coating is another key unit operation of solid dosage form manufacturing and ensures uniform coating films on solid dosage forms required to, e.g., mask bitter tastes of the core formulation, improve stability of the finished product and modify the drug release profile. Coating processes have already been used for several decades in pharmaceutical manufacturing and their design and operating conditions merely relies on art than on fundamental scientific knowledge. In this study we evaluate the effect of variable coating process parameters on coating process and end product quality [2]. Here, the spray coating of a single tablet is simulated, where the tablet passes the spray zone one time at a defined angle. Depending on two coating process parameters, i.e. inlet air temperature and atomizing air pressure, the homogeneity of the coating film thickness is evaluated and an optimal process parameter combination is determined. Data obtained elucidate mechanistic relationships between these input parameters and their effects on coating quality and can form the basis for future studies (e.g., tablets pass the spray zone several times at different angles).

(621c) A Science and Risk-Based Approach to Design Space Drug Product Development

Indra Neil Mukherjee[1], Neil MacPhail[2], Catherine Diimmler[1], Mitchell Colletto[1] and Eric Seymour[1], (1)Merck Manufacturing Division, Merck Sharp & Dohme Corp., West Point, PA, (2)Merck Manufacturing Division, Merck & Co., West Point, PA

Fundamental to the Quality by Design approach is building a high degree of process understanding to ensure critical quality attributes of the final product are consistently met. Providing a foundation for this process understanding comes from a design space, a well defined domain of material and process conditions within which product quality can be predicted and controlled based on identified risks. Based on scientific principles and accounting for risks to quality, a design space was developed at the pilot scale for a fluid bed drug product by using design of experiments (DOE) approach. The use and applicability of this design space upon process scale-up will be discussed.

See more of this Session: TI Application of Quality by Design in Drug Product Process Development
See more of this Group/Topical: Topical i : Comprehensive Quality by Design in Pharmaceutical Development and Manufacture

Spray Drying Process Design: Establishing Links between Process Parameters and Product Performance

Scott Ellis, Ph.D.
San Carlos, CA

Outline

- Introduction – Basis for approach
- Application to design space development
- Examples
  1. Characterization of spray drying in manufacture of a particulate dosage form.
  2. Characterization of spray congealing in manufacture of a tablet dosage form.

Key design elements

1. Drug product design / performance
   - Base specifications on desired product performance and mechanistic understanding.
     - In vitro / in vivo correlations are key.
   - Develop to attain prospectively designed specifications.
   - Define "design space" based on product "critical quality attributes".
2. Manufacturing process design / performance
   - Design to ensure consistent quality and performance.
     - Continuous "real time" assurance of quality based on process performance.
   - Define "control space" as "critical process parameter" subset of design space.
     - Process parameter / in vitro product performance correlations are key.

Inputs to Design Space

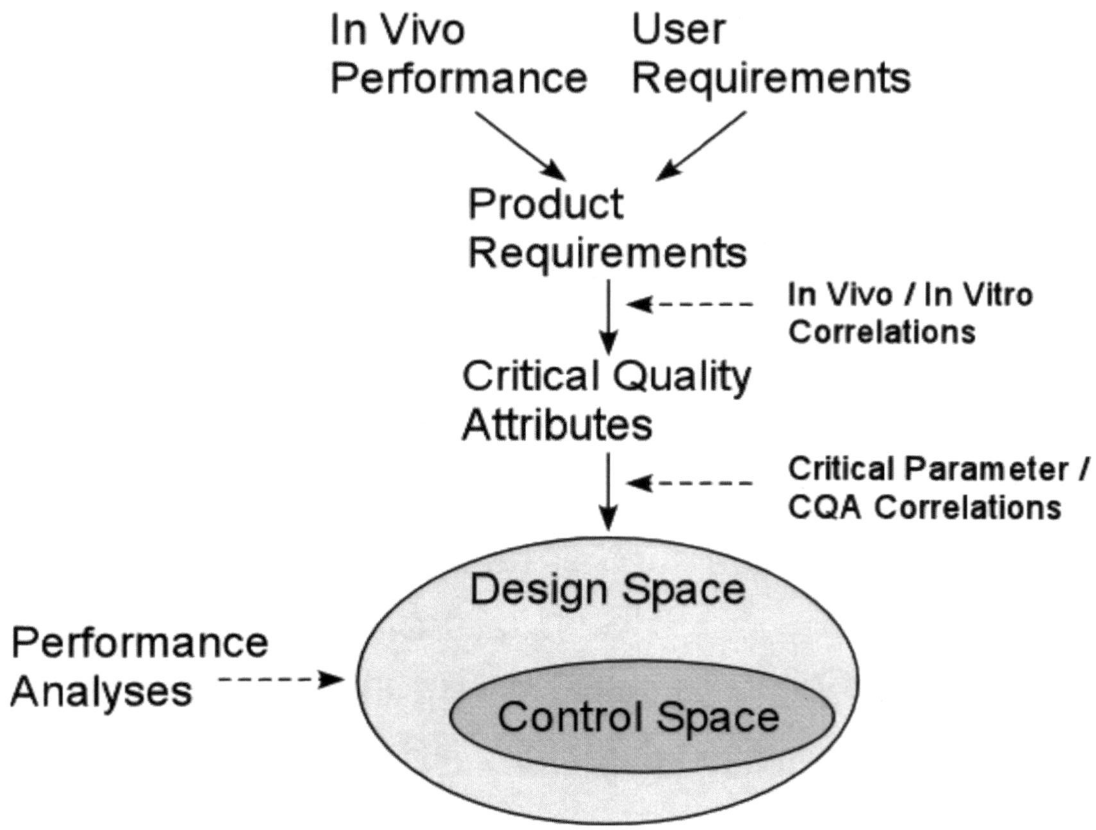

Translating ICH Q8 into Practice

- Quality by Design is a philosophy of product development based on application of underlying science and risk definition / reduction.
  - How to convert this guidance into practice?
- The stepwise path defined by Six Sigma DMEDI (Design for Lean Manufacture) can be used as a logical means of approaching each key product development step.
  - Focuses activities against clinical timeline drivers.
  - Captures tradeoffs as to reducing risk vs. front-loading project.
  - Allows for re-looping to incorporate technology improvements or data-driven changes to product targets.

Systematic Development Path

Objective:

- Identify key dependencies and map design space early in clinical development to reduce late-stage product equivalency (transfer and scale up) risks.

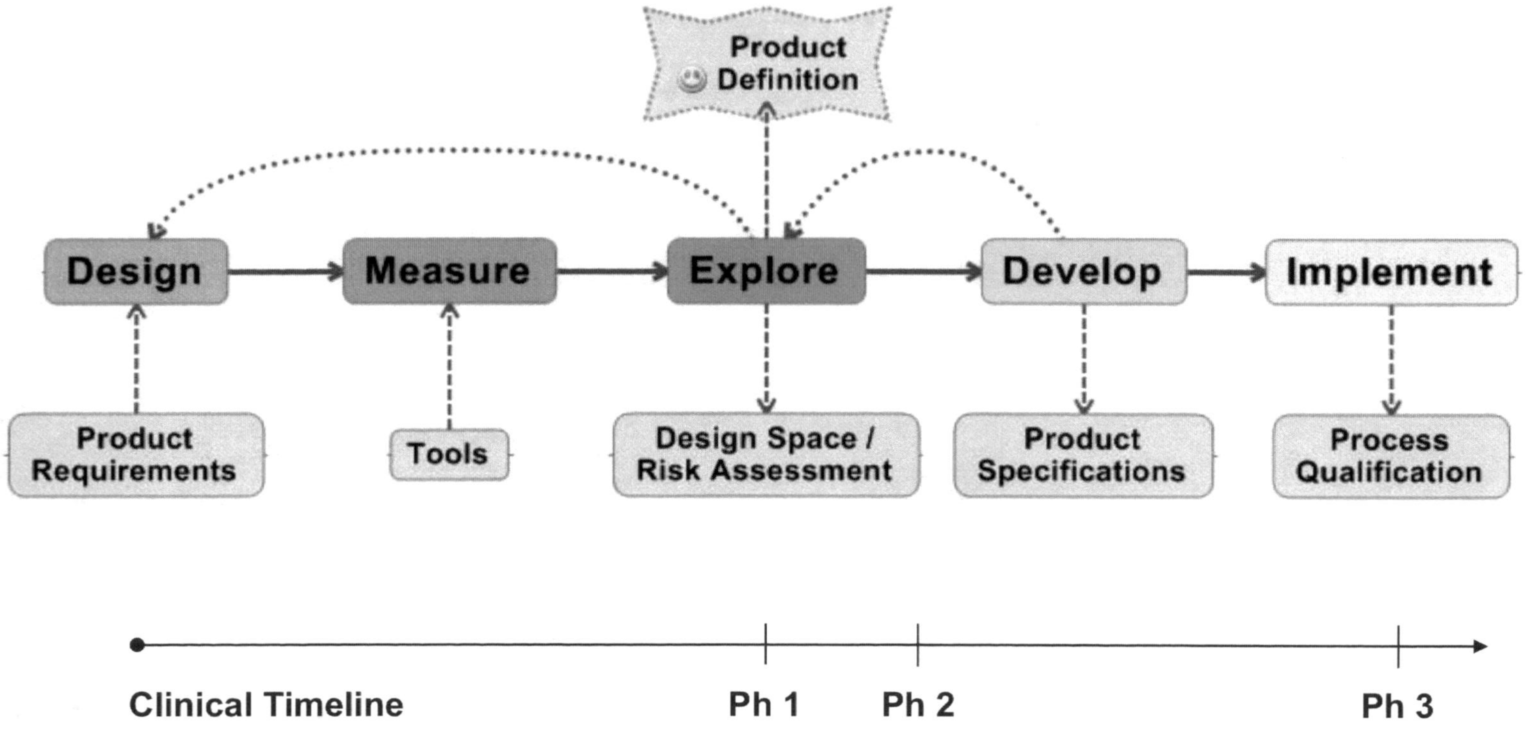

Design Element

Objectives:

- Define target process train based on product requirements, installed equipment base, etc.
- Identify any technology development needs.
- Generate development plan for further refinement.

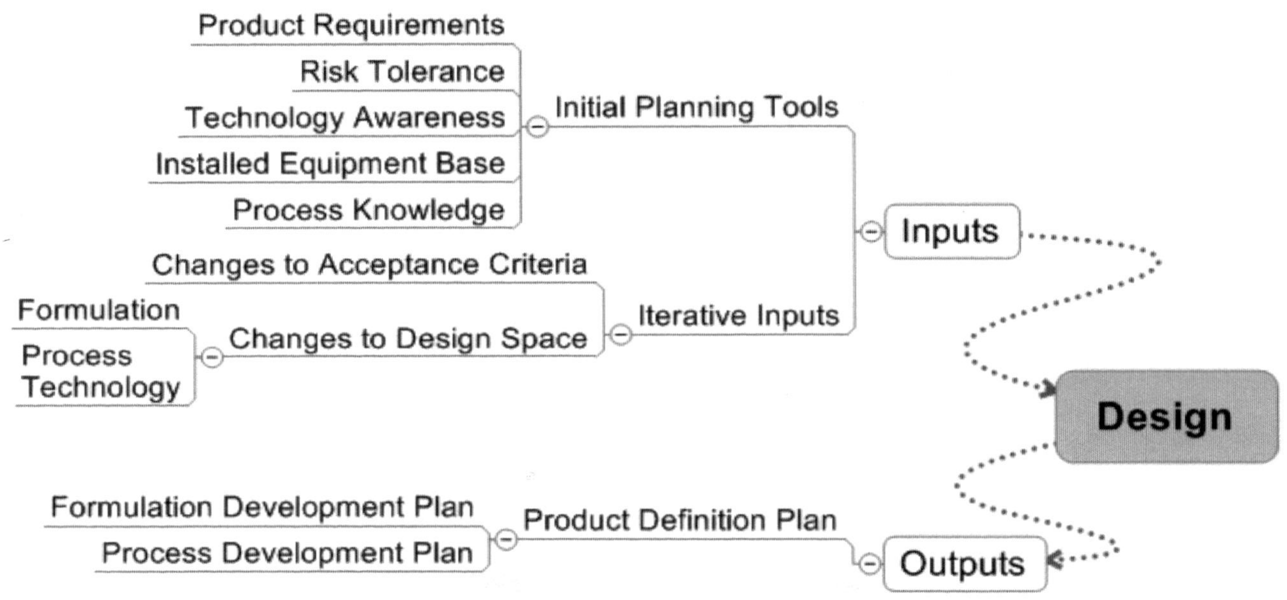

Measurement Element

Objectives:

- Refine study design to capture key parameters, variables, and interactions.
- Generate prototype design space for analysis.
- Categorize tools (e.g. PAT) and desired information.

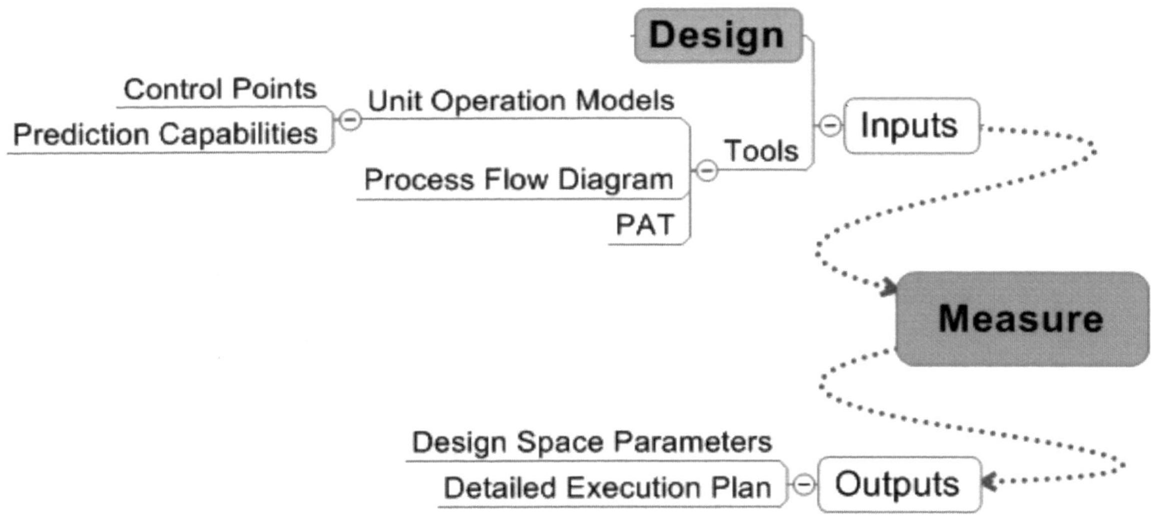

Exploration Element

Objectives:

- Use underlying science, mathematical models, and statistical tools to refine the design space and relate process conditions to product performance.
- Capture process and product failure modes and assess feasibility.

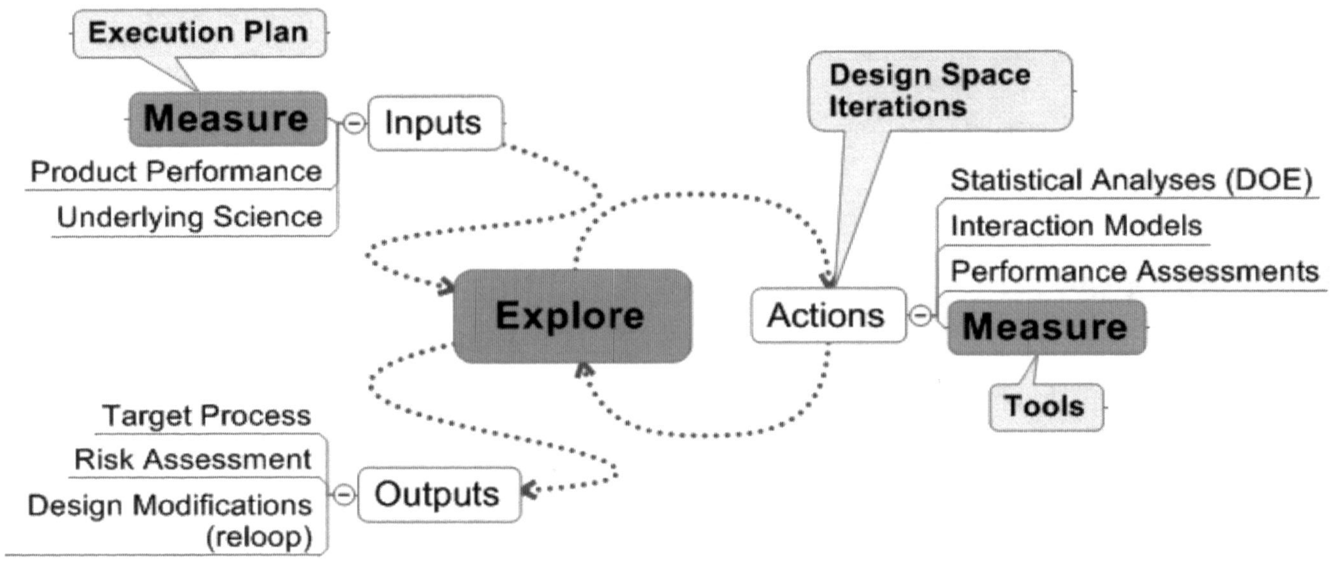

Development Element

Objectives:

- Perform formal FMEA on target process.
- Understand relationships between process reproducibility and product variability (Design for Lean Six Sigma).
- Establish process operating window and product specifications.

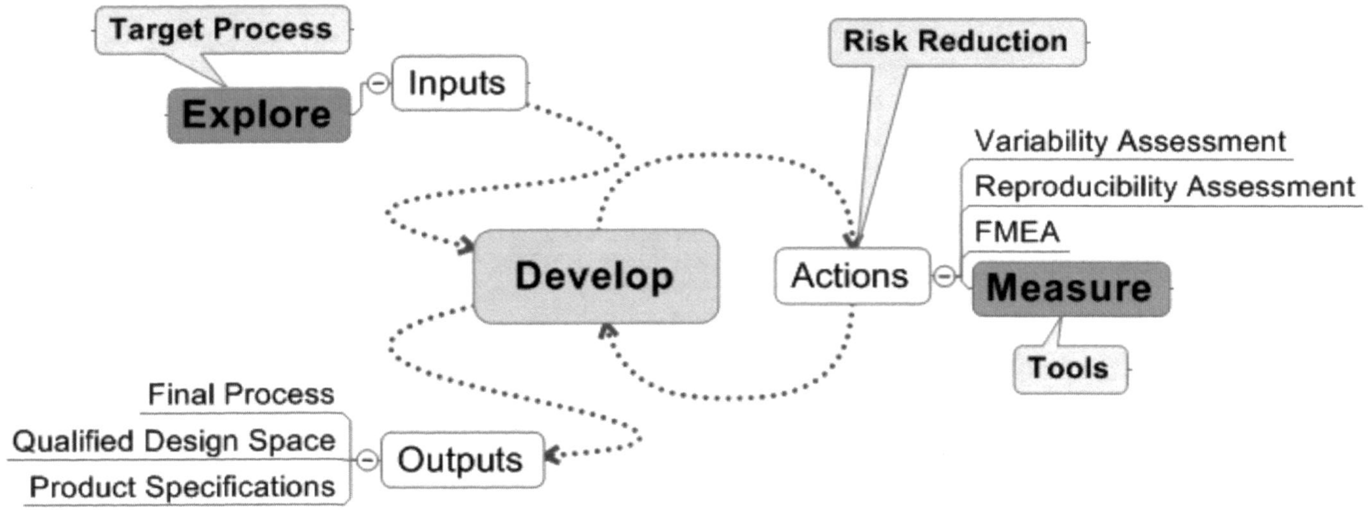

Implementation Element

Objectives:

- Perform formal qualification of final process.
- Initiate any scale up efforts through expansion of design space.
- Note: Assumes that assessments of technology similarity (e.g. scale dependencies) have been addressed *a priori* or in parallel.

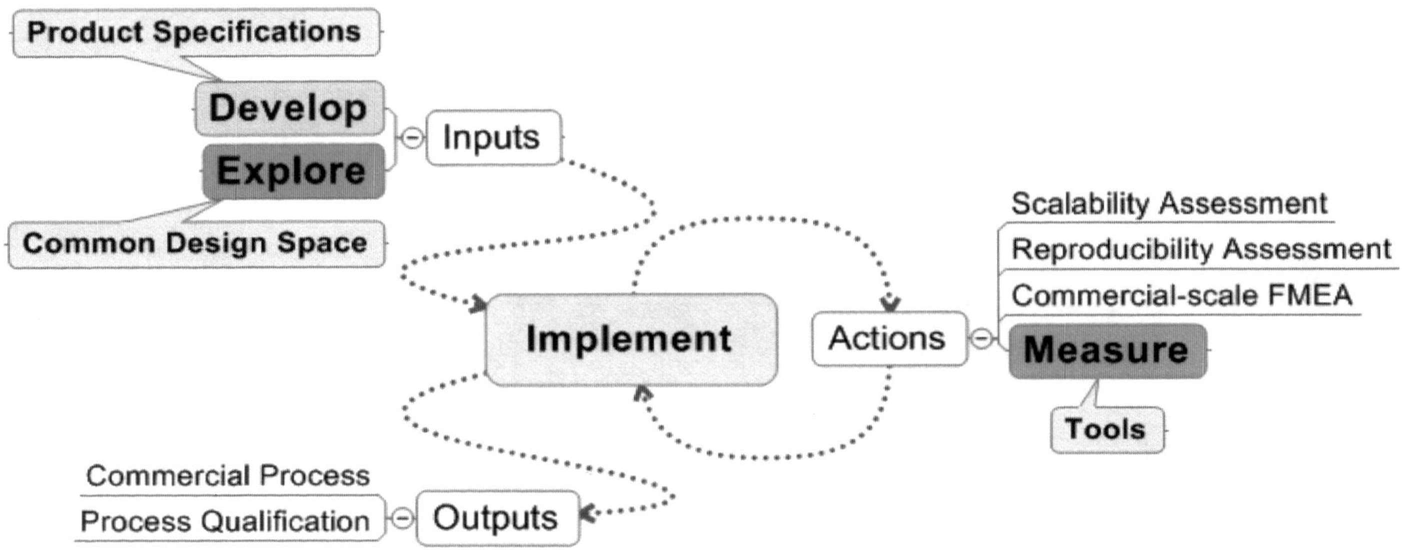

Focus on Exploration

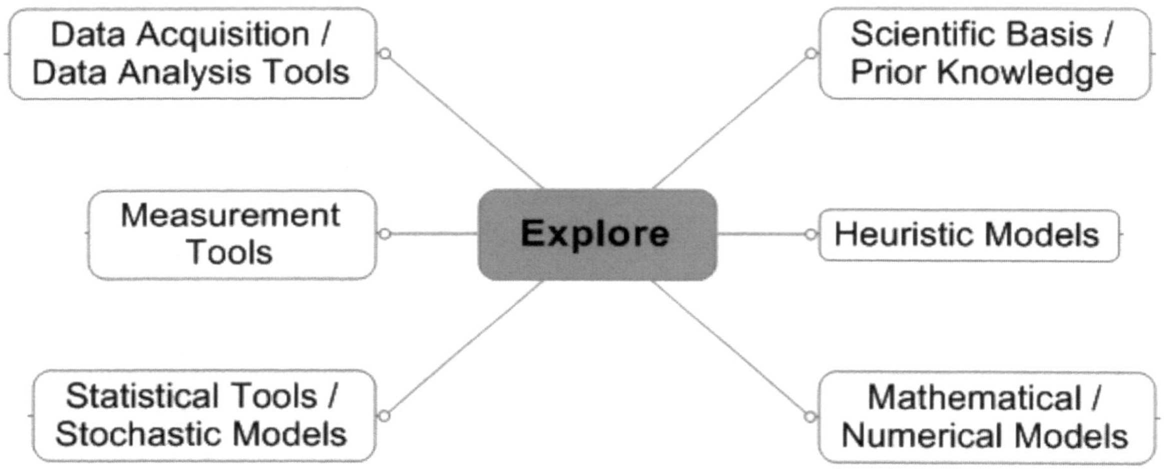

- Design space mapping and product definition occur in this step; outcome enables establishment of risk-based acceptance criteria.
  - Criteria are based on real-time process performance rather than extensive post-process product release testing.
- <u>Challenge</u>: In a process train of many unit operations, how to link product performance back to one or more aspects of each unit operation?

Concept of common design space

- Focus on optimizing product properties, not process conditions (heuristic approach).
- Base the target design space for each unit operation on its realistic operating range as well as on known limitations of the expected product intermediate.
  - "Realistic" includes limits of controllability, API stability, etc.
- Characterize each design space using statistical tools (e.g. experimental designs).
  - Confirm failure modes and capture product property ranges.
- Restrict design spaces of preceding operations by learning from downstream results (backwards from final product).
  - Harmonize inputs and outputs across all processes.
  - Develop "common" design space to facilitate risk-based final product quality assessment.

Example Approaches

1. Spray Drying
    - Coated particle blend for reconstitution
2. Spray Congealing
    - Direct compression tablet

Contribution of Drying / Congealing Process

- Particle generation for further processing
  - Spray Drying: Solvent evaporation (precipitation)
  - Spray Congealing: Melt solidification (freezing or super-cooling)

- Typical particle properties
  - Median size / Size distribution
  - Particle density / Bed density
  - Morphology (surface area / rugosity)
  - Flowability

→ How to relate to performance of final product?
  - Must accommodate impact of both preceding and subsequent processing steps.

Drying / Congealing Process Flow

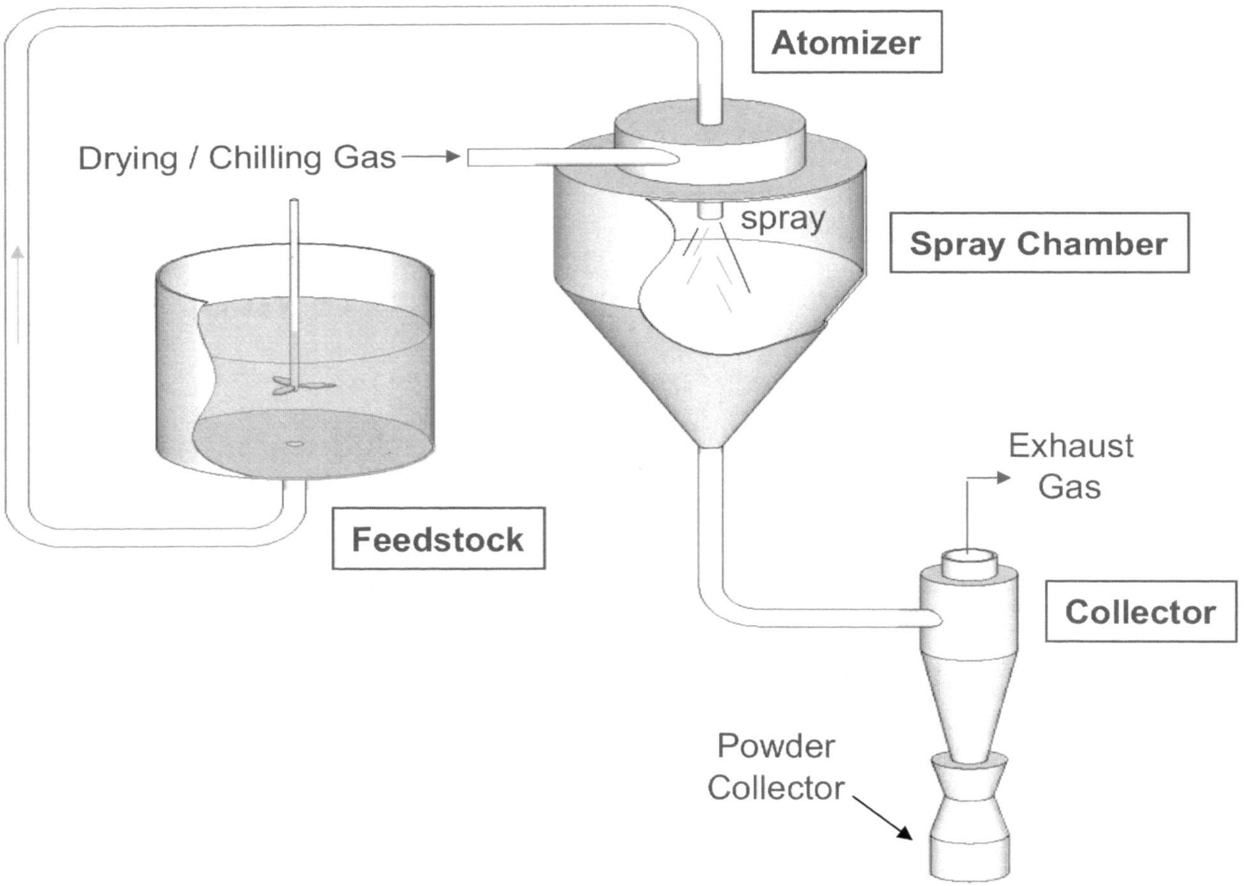

Process Steps

1. Atomization
   - Breakup of liquid feed into discrete droplets.

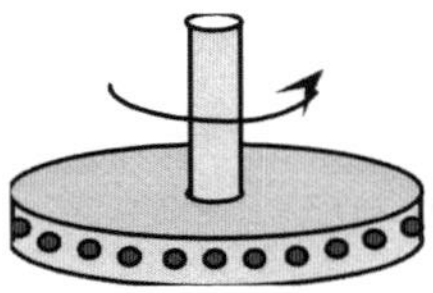

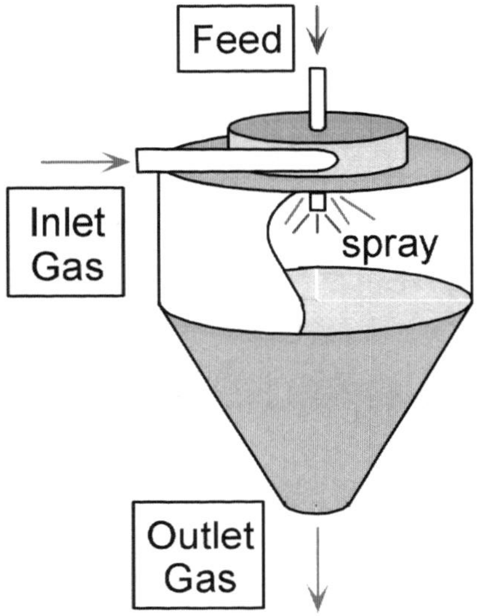

2. Drying / Congealing
   - Transition of droplets to particles.

3. Collection
   - Capture of particles from carrier gas.

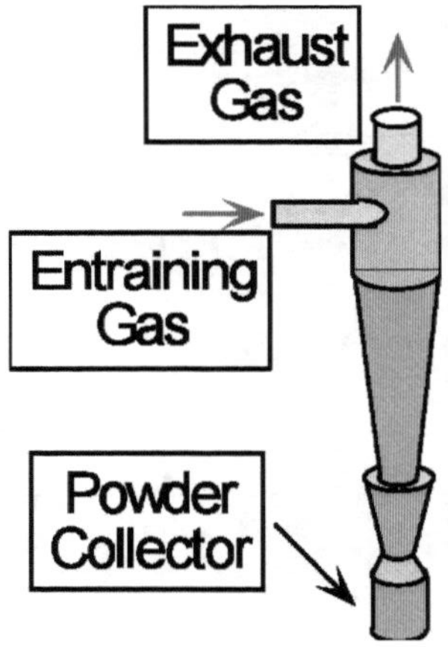

Coated Particle Blend for Reconstitution

- Product is a dry blend of reverse-enteric coated API particles and other excipients, reconstituted at the pharmacy.

- Key product attributes:
  - C_{max} / t_{max} combination (*in vivo* bioavailability)
  - Lack of *in vitro* drug release (in suspension)
  - Taste / mouth feel (very bitter API)
  - Blend uniformity (for filling)

- Role of spray drying:
  - Manufacture of solid drug particles containing ~ 70% ($^{w}/_{w}$) API + bulking agent & buffer.

Coated Particle Blend for Reconstitution

- Rationale for selection of spray drying process:
  - Production of particles with a morphology to facilitate efficient particle coating.
  - Production of small particles with narrow size distribution to provide good mouth feel.
  - Production of particles with median size comparable to other excipients to ensure efficient blending and content uniformity.

- Challenge:
  - To achieve target drug load, particles must be spray dried from an aqueous feedstock containing about 35% ($^{w}/_{w}$) API (as dispersed solid) + 15% binder. API is jet-milled before use. This combination of high feedstock solids content + small API particle size results in a highly thixotropic feedstock. Handling and spraying this feedstock requires additional technology development.

Coated particle blend process flow

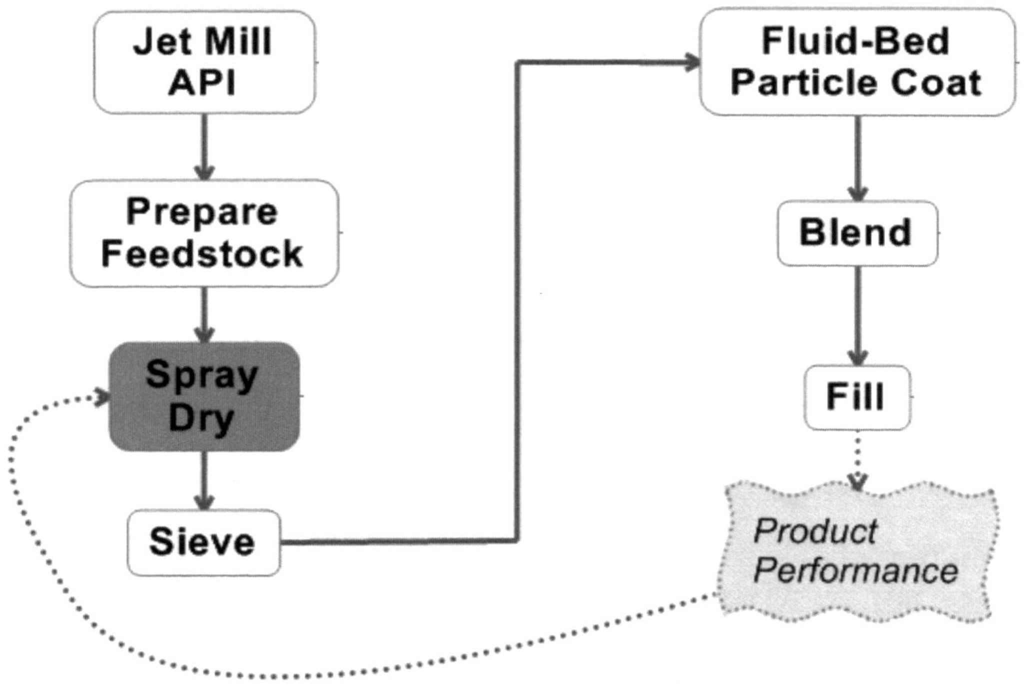

Particle Blend Example: Taste / Mouth Feel

- Objective:
  - Relate taste / mouth feel of reconstituted suspension to performance of spray drying process.

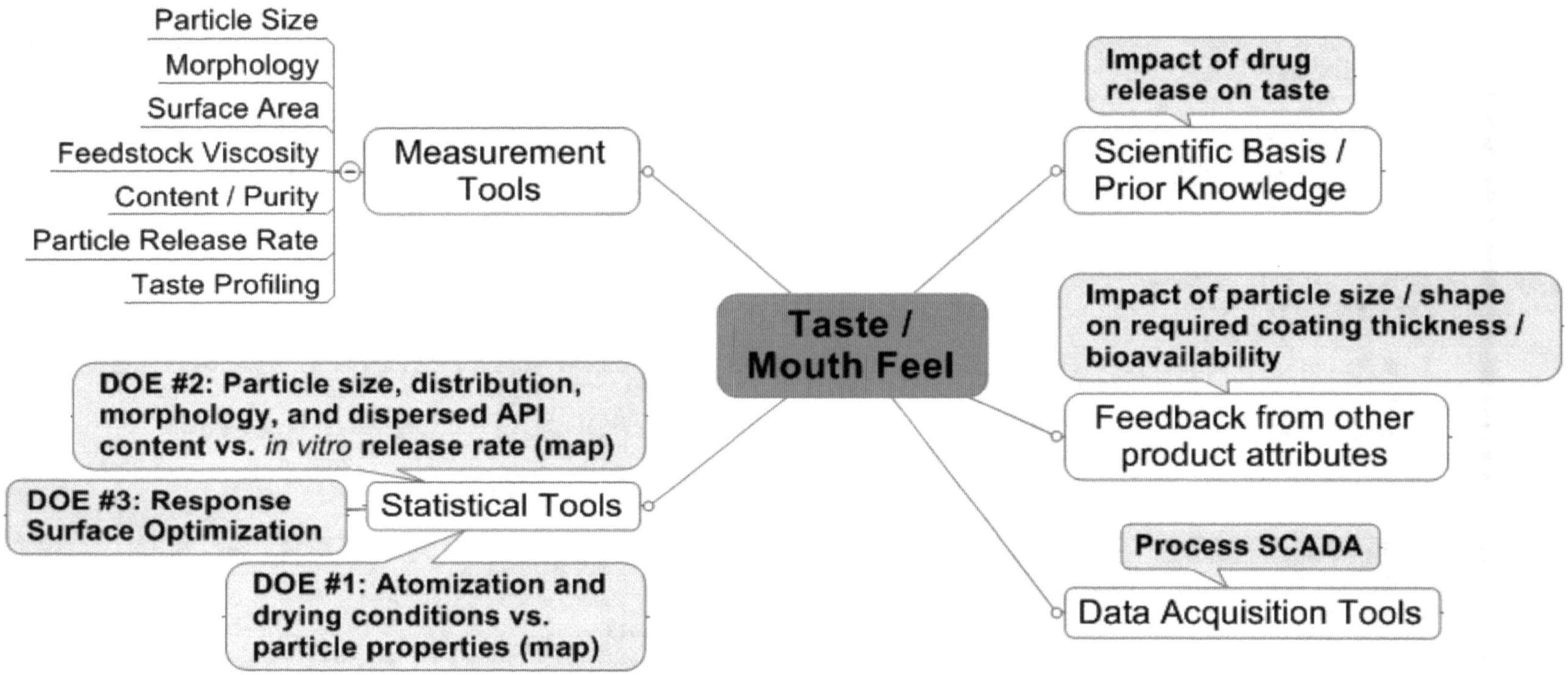

Particle Blend – Results

- Taste and mouth feel requirements were established through taste panel studies (Measurement element).
  - Small, smooth particles were thought to have superior taste (easier to coat uniformly; no thin spots) and mouth feel (not "gritty").
- Design spaces were developed for the high viscosity suspension feedstock and spray drying process:
  - Particle size / distribution / morphology were mapped against feed viscosity (↑ rate → ↓ viscosity), atomizer speed, and drying rate.
  - In vitro release rates were mapped against particle properties for a given coating procedure.
  - Coating process design space was overlapped with spray drying design space to establish common performance window.
    - Acceptable size distribution for fluid bed coating was narrower than the achievable size distribution by spray drying).
  - Desirable minimization of in vitro release rate was reconciled against retention of sufficient bioavailability through common response surface DOE.

Particle Blend – Process Notes

- Key process variables:
  - Atomizer rotational speed
    - Particle size and distribution are reduced at higher speeds.
  - Drying rate
    - Particle morphology becomes less uniform at higher evaporation rates. Tradeoff is between ideal rate and available drying time.
  - Feedstock viscosity / flow rate
    - Control ensures uniform flow to atomizer.
    - Maximize flow rate to minimize viscosity / improve atomization.
- Other notes:
  - In-line viscosity measurements are important in monitoring for particle consistency.
  - Atomizer hardware design / characterization is also key in handling non-Newtonian suspensions – should be captured through a parallel development effort.

Direct Compression Tablet

- Product is a coated tablet made by direct compression (i.e. no granulation / blending).

- Key product attributes:
  - Dissolution rate
  - Hardness / Friability

- Role of spray congealing:
  - Manufacture of solid drug particles containing $< 10\%$ ($^{w}/_{w}$) glidant.

Direct Compression Tablet

- Rationale for selection of spray congealing process:
  - Production of small particles to improve release rate of poorly soluble drug (higher surface area).
  - API is a waxy solid not amenable to traditional milling, granulation, and drying techniques.
  - API can be melted with no loss of activity or stability.

- Challenges:
  - Particle flowability is poor, but may be improved through addition of small amount of glidant as dispersed solid. Glidant must be stable at process temperatures.
  - Handling and spraying a moderate-viscosity (slightly non-Newtonian) melt is uncommon and requires additional technology development.

Direct compression tablet process flow

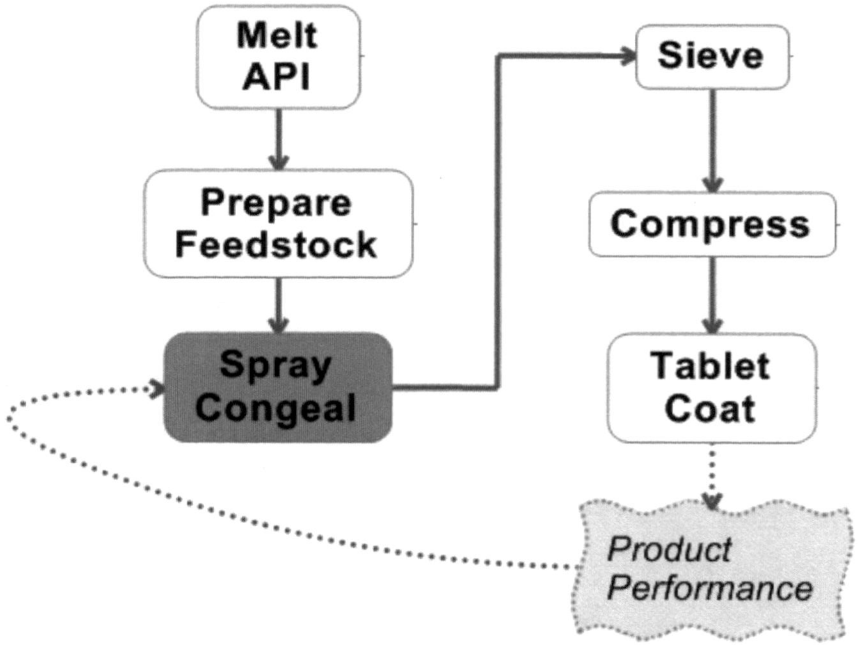

Direct Compression Example: Dissolution Rate

- Objective:
  - Relate target tablet dissolution rate to performance of spray congealing process.

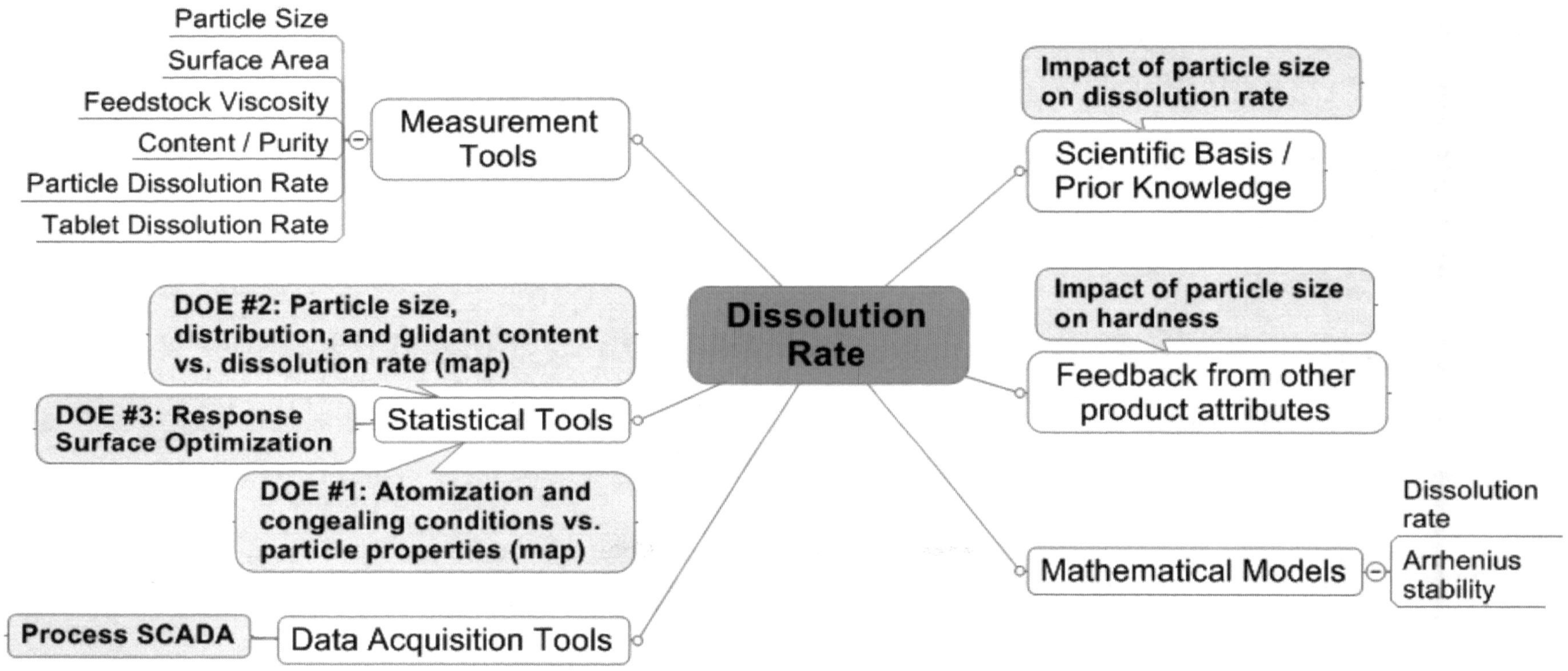

Direct Compression Tablet – Results

- Minimum acceptable drug availability rate was set in Product Requirements exercise (Design element).
  - Enabled backward analysis against target dissolution rate (heuristic approach).
- By evaluating a range of particle sizes, distributions, and glidant contents, a “window” of attributes was determined:
  - Dissolution rate variability was observed to increase with increasing particle size distribution >> defined maximum allowable distribution for a given median size.
  - Dissolution rate increased with decreasing median size and increasing surface area >> minimize median size…
    - but… difficulty of tablet hardness control increased with decreasing median size >> need to re-loop with a third DOE combining all product attributes vs. particle size (response surface optimization).
  - The minimum glidant content needed to facilitate powder flow had no impact on dissolution rate.

Direct Compression Tablet – Process Notes

- DOE on process conditions vs. particle properties
  - Intent was to decouple spray congealing process performance from final product attributes – simplified data analysis.
- Key process variables:
  - Atomizer rotational speed
    - Direct impact on median particle size and distribution.
  - Cooling capacity
    - Complete particle solidification is needed before particles contact vessel walls (or deformation / agglomeration may occur).
  - Feedstock viscosity
    - Control ensures uniform flow to atomizer.
  - Interesting PAT opportunity: Fluctuations in feedstock temperature or glidant content result in small changes in viscosity. This parameter may then be measured in-line to ensure feedstock consistency and product dose content uniformity.

Next Steps

- Use product requirements and design space of next downstream activity to restrict design space of each unit operation.

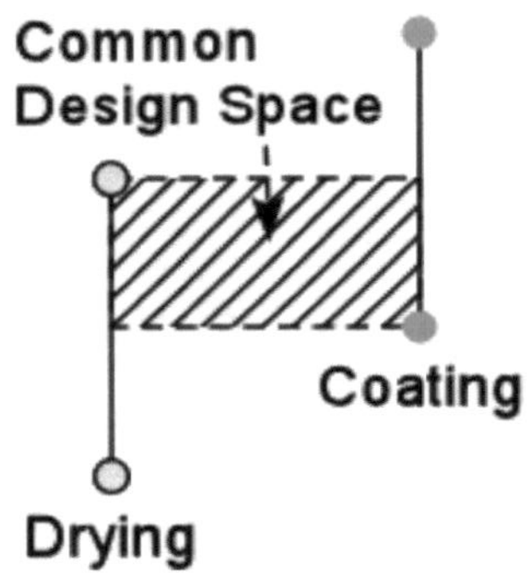

- Optimize product performance through response surface DOE on the common design space.
- Utilize response surface predictions to establish risk-based product acceptance criteria.
- Design for Lean Six Sigma: Work to minimize product variability through continued understanding of operating windows (Development element).

(621e) Drop-On-Demand Dispensing of Pharmaceutical Suspensions: Correlation of Material Attributes and Operating Conditions to Process Performance

Sara Holt, Chemical Engineering, Rutgers University, New Brunswick, NJ, Paul Takhistov, Food Science, Rutgers University, New Brunswick, NJ and Fernando Muzzio, Chemical and Biochemical Engineering, Rutgers University, Piscataway, NJ

The ability to produce drops on demand is critical to microdosing applications, such as dispensing inks, polymers, and other biological fluids. In the pharmaceutical industry microdosing can be used to reliably fill capsules with multiple materials at strictly controlled ratios, under full closed loop control. Additionally the emergent trend of personalized dosing for therapies in hospitals could lead to personalized microbatch manufacturing based on drop on demand microdosing. While the satisfactory operation of microdosing equipment options for different liquids is usually confined to a narrow region of the system parameter space, there is still a significant lack of understanding of the underlying correlations between material attributes, processing parameters and properties of the dispensed droplets.

In this work, a piezo-electric valve, specifically a 50 micron MJ-AT-01 valve from Microfab, has been chosen as a typical micro-dispensing device. The process design space has been mapped for materials with varied properties (viscosity, surface tension, etc.). Process parameters considered include dwell voltage and rising time among others. To better investigate the underlying physical phenomena governing the performance of the system, numerical simulations were performed of the flow of both Newtonian and non-Newtonian fluids through the dispensing nozzle. Specific focus was placed on the resolution of the flow directly before and after the nozzle orifice. The results of this study could provide a substantial contribution to the ability to a priori assess the operating parameters required to reliably dispense drops on demand for different materials.

See more of this Session: TI Application of Quality by Design in Drug Product Process Development
See more of this Group/Topical: Topical i : Comprehensive Quality by Design in Pharmaceutical Development and Manufacture

(621f) Can A Design Space Be Built On Material Attributes Alone?

James N. Michaels[1], Holly Bonsignore[2], Buffy L. Hudson-Curtis[3], Steven Laurenz[4], Homer Lin[5], Thomas Mathai[6], Girish Pande[3], Ashlesh Sheth[7] and Omar Sprockel[8], (1)Merck and Co. Inc., West Point, PA, (2)Pfizer Inc., (3)GlaxoSmithKline, (4)Abbott Laboratories, (5)Amgen, Thousand Oaks, CA, (6)Cephalon, (7)Merck & Co., Inc., (8)Bristol-Myers Squibb Company, New Brunswick, NJ

In April, 2009, the PhRMA Drug Product Technical Group sponsored an industry workshop to explore the practicality and limitations of defining a design space in terms of material attributes rather than process variables. This presentation summarizes and expands on the output of this workshop. A material-attribute design space would be independent of scale and configuration of process equipment and the associated process variables. For this reason, it would be portable in the sense that post-approval changes of equipment scale, nameplate, or location would not require regulatory approval. A hypothetical case study of an immediate release oral tablet is presented to illustrate how such a design space could be constructed and implemented.

See more of this Session: TI Application of Quality by Design in Drug Product Process Development
See more of this Group/Topical: Topical i : Comprehensive Quality by Design in Pharmaceutical Development and Manufacture

(621g) Impact of API Particle Size On a Sustained Release Oral Tablet Formulation

Sarina G. Harris Ma, Derrick Kim, Chen Mao, Laura Maurer and Sami Karaborni, Formulation Development, XenoPort, Inc, Santa Clara, CA

The objective of this study was to understand the impact of API particle size on various blend and tablet quality attributes of a sustained release oral tablet formulation produced via a direct compression process. The API primary particle size and agglomerate size were identified as critical material attributes through a risk assessment performed by preliminary hazards analysis. The two factor, three level experiment design consisted of three lots of API with different primary particle size and agglomerate size controlled by three milling conditions. After milling, the API primary particle size and agglomerate size were characterized by various methods including static light scattering, image analysis, and sieve analysis. Tablets were manufactured at a 1 kg scale using high shear blending followed by direct compression. A model was then developed to assess the impact of API primary and agglomerate particle size on several bulk blend and tablet quality attributes. An approach based on risk assessment and limited design of experiments focused on critical material attributes can generate significant process understanding. The knowledge gained at small scale can be leveraged to generate a preliminary design space.

See more of this Session: TI Application of Quality by Design in Drug Product Process Development
See more of this Group/Topical: Topical i : Comprehensive Quality by Design in Pharmaceutical Development and Manufacture

(673a) A Lab-Scale Investigation Into Drying of An Active Pharmaceutical Ingredient to Gain Process Understanding for Application in Full-Scale Manufacture

Holly Shearer, Pharmaceutical Development, AstraZeneca, Bristol, United Kingdom

Drying of an active pharmaceutical ingredient (API) can often considerably add to the cycle time of a manufacturing process. An investigation has been conducted using a two-litre laboratory pressure filter-dryer in order to gain process understanding to enable the optimisation of drying in the manufacturing process. Water wet API is dried using nitrogen at 40 degrees centigrade in pressure filter-dryers. The effect of batch size and pressure within the filter-dryer on overall drying time was investigated in the small-scale rig. Laboratory results were compared to data from the manufacturing plant to determine their relevance to the manufacturing process. A humidity probe was positioned in the outlet gas of the laboratory filter dryer. Data collected was correlated with loss-on-drying analysis of samples of the product to determine the humidity of the gas that indicated the product was dry. The instant feedback provided by the humidity probe also enabled the use of the agitator during drying to be optimised.

See more of this Session: TI Application of Quality by Design In API Process Development
See more of this Group/Topical: Topical i : Comprehensive Quality by Design in Pharmaceutical Development and Manufacture

(673b) Pragmatic Application of Crystallization Design Basics for a Pharmaceutical Compound

James Wertman, Megan Ackers and Philip Dell'Orco, GlaxoSmithKline, Upper Merion, PA

An API crystallization can be sufficiently developed to provide a robust process utilizing basic design principles and simplified modeling techniques. Such an approach is presented as an example of the typical methods for an industrial application, where resource and time are becoming increasingly important.

See more of this Session: TI Application of Quality by Design In API Process Development
See more of this Group/Topical: Topical i : Comprehensive Quality by Design in Pharmaceutical Development and Manufacture

(673c) A Risk-Based Approach to the Application of Biocatalysis for Manufacture of Small Molecule APIs

Stuart Goodall[1], **John Wong**[2], Gregory Finch[2], Timothy J. Watson[3], Deborah Finco[2] and Ned Mozier[4], (1)Pfizer Inc, Sandwich, United Kingdom, (2)Pfizer Inc, Groton, CT, (3)Pfizer, Inc, Groton, CT, (4) Pfizer Inc, Chesterfield, MO

Enzymes are being routinely used for synthesis of organic molecules including small molecule APIs as they offer significant benefits (e.g. cheaper, greener, more selective, higher yielding, safer etc) over conventional chemo-catalytic processses. Potential impurities that could occur in products of enzymatic synthesis depend on several factors such as the enzyme source (plant, animal, microbial), the manufacturing process for the enzyme, and the enzymatic reaction step, and include proteins (enzyme and host cell), nucleic acids, endotoxins, and bioburden. Current regulatory guidance does not adequately address enzyme related impurities for small molecule APIs and this lack of clear guidance has been reflected in regulatory queries received for a number of Pfizer enzymatic regulatory submissions. The presentation will highlight the risk assessment framework, built around Quality By Design and drug safety principles, used within Pfizer and present case studies from two recent commercial filings

See more of this Session: TI Application of Quality by Design In API Process Development
See more of this Group/Topical: Topical i : Comprehensive Quality by Design in Pharmaceutical Development and Manufacture

(673d) Modeling Approach for the Scale-up of a Reactive Distillation System

Isabel Figueroa and Shekhar Viswanath, Eli Lilly and Company, Indianapolis, IN

Reactive distillations are a widely used unit operation in reactive systems that are governed by a chemical equilibrium such as esterification and ester hydrolysis reactions. The process performance often times is quantified by the distillation time and/or the reaction final yield. These are results of the coupling of phenomena that are scale-dependent (heat transfer in boiling regime, mass transfer limitations) as well as scale-independent processes (e.g.: reaction thermodynamics, reaction kinetics, calorimetry). Therefore, predicting the process performance based solely on small-scale experimentation may lead to erroneous estimations and misunderstanding of how the process variables interplay. In this work, we present a scale-up approach based on a combination of first principles modeling and experimentation. We estimate the process performance at different scales and study the sensitivity of the process to operational parameters such as heat transfer driving force, solvent recycle, removed fraction of volatile, etc. Our approach is capable of robustly predicting process outcomes at lab and pilot-plant scale and delivers a better understanding of the underlying physics governing this kind of processes.

See more of this Session: TI Application of Quality by Design In API Process Development
See more of this Group/Topical: Topical i : Comprehensive Quality by Design in Pharmaceutical Development and Manufacture

(673e) Sequential Design of Experiments for Design Space Definition of the Crystallization of An Active Pharmaceutical Ingredient

Eric Sirota[1], James P. Corry[1], Ralph Calabria[1], Paul James[2], Shane Grosser[1], Luke Schenck[1], Aaron Cote[1] and Aaron Moment[1], (1)Chemical Process Development and Commercialization, Merck & Co., Rahway, NJ, (2)Center for Mathematical Sciences, Merck & Co., Rahway, NJ

A sequential design of experiments performed at lab scale was used to define the factory design space of an API semi-continuous reactive crystallization. Following risk assessment activities and exploratory experiments, a fractionated design was run to determine statistically significant factors and their interactions. When the contributions of non-statistically significant factors were removed from the model, this fractional DoE collapsed to a full factorial. Curvature in the key responses necessitated further augmenting the design with custom axial points for response surface mapping. This resulted in an empirical model that predicted particle size and facilitated design space definition for the crystallization.

Also discussed will be the design of the reactive crystallization, which employed media milled seed and sonication to reach a desired, unimodal particle size distribution without use of dry milling. The performance of the crystallization process across lab, pilot plant, and factory scale will be discussed.

See more of this Session: TI Application of Quality by Design In API Process Development
See more of this Group/Topical: Topical i : Comprehensive Quality by Design in Pharmaceutical Development and Manufacture

(673f) Plantwide Dynamic Simulation and Model-Based Control of a Continuous Pharmaceutical Process

Dimitrios I. Gerogiorgis, Department of Chemical Engineering, Massachusetts Institute of Technology (M.I.T.), Cambridge, MA

ABSTRACT

Continuous Pharmaceutical Manufacturing (CPM) constitutes a groundbreaking paradigm in modern pharmacopeia: multinational pharmaceutical industries face unprecedented profitability gaps, due to decreasing drug prices, increasing Research & Development (R&D) and operating expenditures. These business challenges emerge in the era of globalization and can be attributed predominantly (but not exclusively) to the transformation of patent legislation, the vast reduction of patent life, the harmonization of regulatory specifications and the advent of generics; new plants must rely on optimization, standardization and modularization ideas (Behr et al., 2004).

Batch pharmaceutical processes are widespread, forming the overwhelming majority of installed manufacturing plants; nevertheless, most are based on suboptimal recipes focused on conforming with the strict licensing requirements for drug product purity, stability and efficacy (ICH, 2009). *Continuous pharmaceutical processes*, though, need rely on systematic (not empirical) process understanding and R&D, therefore necessitating the employment of process modeling towards the articulation of *Design Space* (Yu, 2008) and the use of *Process Analytical Technology (PAT)* (Munson, 2006); this may turn out a time-consuming endeavor for novel manufacturing lines. The U.S. Food & Drug Administration (FDA) thus spearheads the pursuit of CPM initiatives, recognizing their potential to reduce operating cost at high quality (FDA, 2008) as confirmed in comparative technical (Betz et al., 2003) and modeling (Gerogiorgis & Barton, 2009) studies.

Modern batch manufacturing processes are frequently operated at maximum unit efficiency via dynamic optimization models, relying on mathematical rigor to capture critical phenomena, empirically or mechanistically (Chekhova et al., 2000). However, process scale-up hinders model adaptation. Optimal external input profiles can be computed and applied *during* each batch processed (low-level regulatory control), and statistical methods for offline quality assessment (QA) can reveal deviation trends (high-level supervisory control). Product quality and uniformity are evaluated *a posteriori*: often, drug substance (DS) or drug product (DS) lots must be discarded due to their poor Critical Quality Attributes (CQA). The characteristic batch plant time for a given lot size is the sum of processing and idle times required for its production.

Continuous manufacturing lines, in contrast to batch tank farms, eliminate idle times between processing units by definition; however, deeper understanding is required for their operation. Plantwide dynamic model building requires significant effort and parameter estimation implies arduous, costly experiments which are not a likely luxury if rapid launching is of the essence. The characteristic timescale for a novel CPM plant of given capacity is harder to define, since it depends on structural complexity, unit interdependence, recycles, rate data and timescale spectra. Benign (dilute) or challenging (dense) flow chemistries and multicomponent mixtures affect reaction and separation times, as pointed out in a CPM miniplant study (Roberge et al., 2008). Online sensing and manipulation is possible via modern PAT sensors and actuators, but empirical models can neither explain nor predict how local disturbances may affect plantwide quality performance (Ward et al., 2010), as fundamental phenomena are often poorly understood. CPM plantwide dynamics, therefore, remain a hitherto elusive (hence challenging) objective.

The present study focuses on plantwide dynamics and control of a novel upstream (DS) process which is aimed at the continuous production of a novel Active Pharmaceutical Ingredient (API). The CPM flowsheet comprises 3 organic synthesis reaction and 4 separation steps; individual dynamic models for all respective unit operations are proposed on the basis of experimental data. Open-loop plantwide process simulation in MATLAB (Simulink) allows the determination of timescales, illustrating how common feed disturbances propagate to CQA (e.g. product quality). Control loop pairings can be identified by implementing a RGA method (Skogestad, 2005) and closed-loop plantwide process simulations provide insight into their efficiency and robustness.

REFERENCES

Behr, A., Brehme, V.A., Ewers, C.L., Grön, H., Kimmel, T., Küppers, S., Symietz, I., New developments in chemical engineering for the production of drug substances. *Eng. Life Sci.* **4**(1): 15-24 (2004).

Betz, G., Junker-Purgin, P., Leuenberger, H., Batch and continuous processing in the production of pharmaceutical granules. *Pharm. Dev. Technol.* **8**(3): 289-297 (2003).

Chekhova, E., Barton, P.I., Gorak, A., Optimal operation processes of discrete-continuous biochemical processes. *Comput. Chem. Eng.* **24**(2-7): 1167-1173 (2000).

European Medicines Agency (EMEA), *Impurities: Residual solvents*, VICH Topic GL18 (plus annexes), http://www.emea.europa.eu/pdfs/vet/vich/050299en.pdf (2000).

Gerogiorgis, D.I., Barton, P.I., Steady state optimization of a continuous pharmaceutical process. *Computer-Aided Chemical Engineering (Proceedings of PSE 2009)* **27**(1): 927-932, Elsevier, Amsterdam (2009).

International Conference on Harmonisation of Technical Requirements for Registration of Pharmaceuticals (ICH), *Guidance Documents Q8 (Pharmaceutical Development), Q9 (Quality Risk*

Management), Q10 (Pharmaceutical Quality System), http://www.ich.org (2010).

Munson, J., Stanfield, C.F., Gujral, B., A review of process analytical technology (PAT) in the US pharmaceutical industry. *Curr. Pharm. Anal.* **2**(4): 405-414 (2006).

Roberge, D.M., Zimmermann, B., Rainone, F., Gottsponer, M., Eyholzer, M., Kockmann, N., Microreactor technology and continuous processes in the fine chemical and pharmaceutical industry: Is the revolution underway?. *Org. Proc. Res. Dev.* **12**(5): 905-910 (2008).

Skogestad, S., Postlethwaite I., *Multivariable feedback control: Analysis & design.* Wiley, Chichester (2005).

U.S. Food and Drug Administration (FDA-CDER), *Innovation & continuous improvement in pharmaceutical manufacturing*, www.fda.gov/cder/gmp/gmp2004/manufSciWP.pdf (2008).

Ward, J.D., Yu,C.-C., Doherty, M.F., Plantwide dynamics and control of processes with crystallization, *Comput. Chem. Eng.* **34**(1): 112-121 (2010).

Yu, L.X., Pharmaceutical Quality by Design: Product and process development, understanding, and control. *Pharm. Res.* **25**(4): 781-791 (2008).

See more of this Session: TI Application of Quality by Design In API Process Development
See more of this Group/Topical: Topical i : Comprehensive Quality by Design in Pharmaceutical Development and Manufacture

(2010 Annual Meeting)

(673g) Quality Control by QbD-Guided Optimization of Reaction and Crystallization

Evan T. Barlow, Frederic Buono, Thomas LaPorte, Jaan Pesti and Lori Spangler, Process R&D, Bristol-Myers Squibb, New Brunswick, NJ

Under the QbD paradigm, costs of rework and inspection are minimized by assuring product quality based on control of critical process parameter variation. The case study presented highlights an approach used to identify, quantify, and manage / control risk to product quality by: (1) conducting a Failure-Mode Events Analysis (FMEA) of the process as a quality risk assessment, (2) quantifying risk by determining the impact of multivariate parameter variation on product quality, and (3) developing a control strategy to manage risk to product quality based on scientific understanding of the multivariate design space.

This approach was applied to the manufacture of an advanced intermediate product. The step involves a chlorination reaction, a coupling reactive crystallization, and a reactive recrystallization. The process FMEA highlighted parameter variation for both the chlorination reaction and the reactive recrystallization as potential risks. Experimental design around the chlorination reaction revealed slight increases in a key reagent resulted in lower quality intermediate and much slower reaction kinetics. Product crystallization during the coupling reaction results in substantial residual intermediate entrapment within the product crystals. A 2^3 full-factorial experimental design was implemented to map the reactive recrystallization design space. These and other experiments provided an understanding of the fate of residual intermediate in the reactive recrystallization. Control strategies for both the chlorination reaction and reactive recrystallization were developed to ensure product quality. The effectiveness of the control strategy was demonstrated at manufacturing scale to produce high quality product.

See more of this Session: TI Application of Quality by Design In API Process Development
See more of this Group/Topical: Topical i : Comprehensive Quality by Design in Pharmaceutical Development and Manufacture

(697a) Computational Predictive Models Applied to Scale-up of Solid Oral Drug Products

Mary T. am Ende[1], Rahul Bharadwaj[1], Pankaj Doshi[2], Salvador Garcia-Munoz[1], William R. Ketterhagen[1] and Andrew Prpich[1], (1)Process Modeling & Engineering Technology, Pfizer Global Research & Development, Groton, CT, (2)Chemical Engineering and Process Division, National Chemical Laboratory, Pune, India

The budget-restricted pharmaceutical environment is countered by the heightened expectations for drug products to be developed using science-based principles. These two opposing factors have created a crucial opportunity for engineering principles to be applied and implemented across the industry. In this presentation, modeling techniques applied to formulation and processing operations are discussed as support to the design, development, and scale-up for solid oral drug products. These process modeling techniques are discussed and exemplified with case studies ranging from raw material specifications and formulation property predictions to process parameter predictions. The major processing to produce tablets include i) powder processing; ii) dry and wet granulation; iii) tablet compression; and iv) film coating. Specific consideration has been given to transfer and scale-up issues along with general process design related challenges to pharmaceutical process R&D.

See more of this Session: Model Based QbD in Drug Substance or Drug Product
See more of this Group/Topical: Topical i : Comprehensive Quality by Design in Pharmaceutical Development and Manufacture

(697b) Achieving QM Accuracy at a Fraction of Computational Cost: Lattice Energy Minimization of Organic Crystals Containing Highly Flexible Molecules

Andrei V. Kazantsev[1], Panagiotis Karamertzanis[2], Claire S. Adjiman[1] and Constantinos C. Pantelides[3], (1)Department of Chemical Engineering, Imperial College London, London, United Kingdom, (2)Centre for Process Systems Engineering, Imperial College London, London, United Kingdom, (3)Process Systems Enterprise Ltd., London, United Kingdom

There is a multitude of possible formulation pathways for pharmaceutical compounds among which polymorph, solvate, salt and cocrystal formation are the most common. Finding the optimal formulation by experimentation alone is often expensive and time consuming. In a quality by design environment, computational approaches can aid and direct the search for the best crystal form.

We have recently developed a novel algorithm for the accurate minimization of lattice energy of crystals involving flexible organic molecules [1]. The algorithm makes use of multipole moments (up to hexadecapole level) for the calculation of the intermolecular electrostatic interactions. It derives its accuracy from the use of isolated-molecule quantum mechanical calculations for the computation of intramolecular energy and the electrostatic field around the molecule.

The main novelty of the algorithm is the use of dynamically constructed and updated local approximate models (LAMs) which essentially make available the full accuracy of the quantum mechanical model at each and every iteration of the minimization algorithm while requiring the performance of only a small number (typically 2-8) of quantum mechanical calculations. This has made possible for the first time the accurate treatment of molecules involving relatively large numbers of atoms with significant flexibility in torsional and bond angles and even bond lengths. The algorithm can handle systems with different molecules or chemical entities in the unit cell. In this paper, we discuss its use for the minimization of the lattice energy of crystals involving salts, solvates and cocrystals. The applicability of the approach in ab initio crystal structure prediction studies is also demonstrated.

[1] Kazantsev, A.V. et al.; "CrystalOptimizer: An Efficient Algorithm for Lattice Energy Minimisation of Organic Crystals Using Isolated-Molecule Quantum Mechanical Calculations" in Process Systems Engineering. Volume 6: Molecular Systems Engineering; Adjiman C. S. and Galindo A. (Eds.); Wiley-VCH, Hamburg, 2010

See more of this Session: Model Based QbD in Drug Substance or Drug Product
See more of this Group/Topical: Topical i : Comprehensive Quality by Design in Pharmaceutical

(697c) Applications of Model-Based Quality by Design for Reaction Engineering

Stephen A. Guzikowski, James S. Bergum, Michael P. Cassidy, Chiajen J. Lai, Dong Lin, Sunil S. Patel, Michael E. Randazzo, Thomas M. Razler, Emily A. Reiff, Victor W. Rosso, Jose Tabora, John E. Thornton and Xuejun Xu, Bristol-Myers Squibb, New Brunswick, NJ

Chemical process development implementing Quality by Design (QbD) methodology requires a comprehensive risk assessment, understanding of process parameter multivariate interactions, establishing a potential design space, and ultimately the implementation of a robust control strategy to ensure quality.

In this case study, a multifactor experimental design protocol was employed utilizing automated reactor blocks with robotic powder dispensing, liquid handling, and sampling capabilities to elucidate the effect of reaction parameters on reaction rate and reaction selectivity. The Design of Experiments (DoE) output was used to generate empirical reaction kinetics models for the desired and undesired reaction pathway which were subsequently used to perform quantitative recursive risk assessment and define acceptable ranges for operational flexibility and a potential design space. A normal operating range (NOR) was established by refining reaction parameters which provided a predicted response within tolerance levels while affording adequate practical parameter flexibility. A design plan demonstrating variability within the normal operating range was adopted at pilot scale to demonstrate process scale independence, process robustness and adequate process control.

We will present an expansion of this procedure to a series of synthetic steps demonstrating the utility of this approach to predict and control reaction parameters on scale.

See more of this Session: Model Based QbD in Drug Substance or Drug Product
See more of this Group/Topical: Topical i : Comprehensive Quality by Design in Pharmaceutical Development and Manufacture

Mixture Design and Design Space

Pat Whitcomb
Stat-Ease, Inc.
612.746.2036
fax 612.746.2056
pat@statease.com

Mark Anderson
Stat-Ease, Inc.

2010 AIChE Annual Meeting

Agenda

- **Mixture *(hydrophilic tablet)***
  - Build, run and analyze
  - Numeric Optimization
  - Design Space Determination

Tablet Formulation
Mixture Review

Let's look at a case study involving five components used to form a hydrophilic tablet. The goal is to formulate a sustained release tablet based on a hydrophilic polymer. The polymer swells in the presence of water and impedes drug release. *(Drug is released by a combination of diffusion and erosion.)*

We want to examine the effect of changing the proportions of polymer, drug and three excipients on four responses.

All other components (e.g. lubricant) are fixed.

Tablet Formulation
What was Varied

Here are the five components (and constraints) for our hydrophilic tablet mixture design:

5% ≤ **Lactose** ≤ **42%**

5% ≤ calcium **Phosphate** ≤ **47%**

5% ≤ microcrystalline **Cellulose** ≤ **52%**

17% ≤ hydrophilic **Polymer** ≤**25%**

1% ≤ **Drug** ≤ **2%**

Total = 100 wt %

Tablet Formulation
What was Measured

Responses:

- t(50%) – the time measured in hours for 50% drug release.
- shape – a factor describing the release profile:
  - shape = $(t_{75\%} - t_{25\%})/t_{50\%}$
  - shape = 1.0 *zero order; constant release*
  - shape = 1.5 *first order; proportional release*
- hardness kP (kiloponds)
- friability % – measure of crumbliness.

Agenda

- Mixture *(hydrophilic tablet)*
  - **Build, run and analyze**
  - Numeric Optimization
  - Design Space Determination

Tablet Formulation
Constrained Mixture Design *(page 1 of 2)*

1. Build a 5 component optimal mixture design:

Name	Low	High
Lactose	5	42
Phosphate	5	47
Cellulose	5	52
Polymer	17	25
Drug	1	2
Total = 100 wt %		

Tablet Formulation
Constrained Mixture Design *(page 2 of 2)*

2. Design for a quadratic mixture model using integrated variance (IV) as the optimality criteria. Pick:
   - 15 model points using the optimality criteria.
   - 5 lack of fit points using distance as the criteria.
   - 5 replicates using the optimality criteria.
3. There are four responses:

Name	Units
t(50%)	h
shape	factor
hardness	kP
friability	%

Tablet Formulation
Design (25 runs) with Results

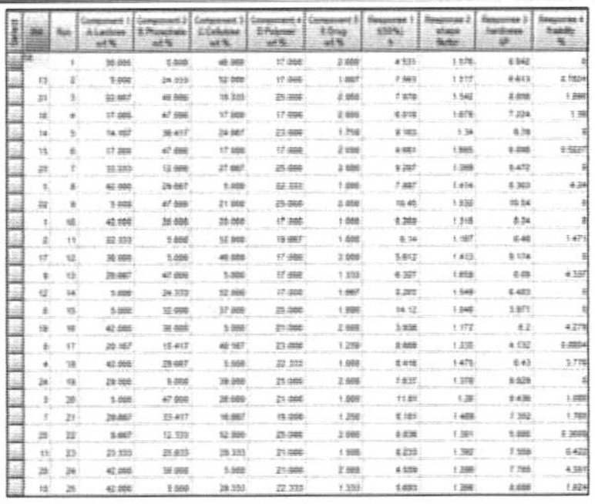

Tablet Formulation
Model Summary

Response	Model p-value	LoF p-value	Adjusted R^2	Predicted R^2	Adequate precision
t(50%)	< 0.0001	0.6511	0.9373	0.8671	24.59
shape	< 0.0001	0.7113	0.8748	0.7580	15.84
hardness	0.0001	0.7529	0.7788	0.5647	12.04
friability	< 0.0001	0.8589	0.9566	0.8647	17.35

Responses are modeled with a quadratic mixture that is reduced by backward selection to remove terms with p-values > 0.1000.

Agenda

- Mixture *(hydrophilic tablet)*
  - Build, run and analyze
  - **Numeric Optimization**
  - Design Space Determination

Tablet Formulation
Numeric Optimization

Numeric optimization criteria:

Response	Goal	Lower	Higher
t(50%)	target = 8	6	10
shape	minimize	1	1.5
hardness	target = 6	4	9
friability	minimize	1	2

Optimize for tablets with two drug dosages:
- Regular strength drug = 1 wt%
- Extra strength drug = 2 wt%.

Tablet Formulation
Drug = 1%

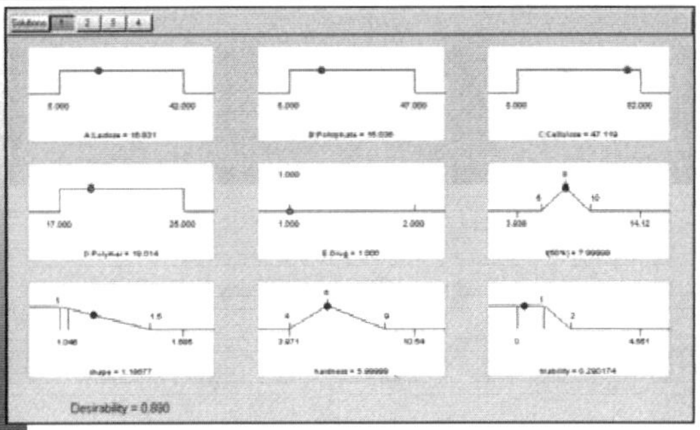

Tablet Formulation
Drug = 1%

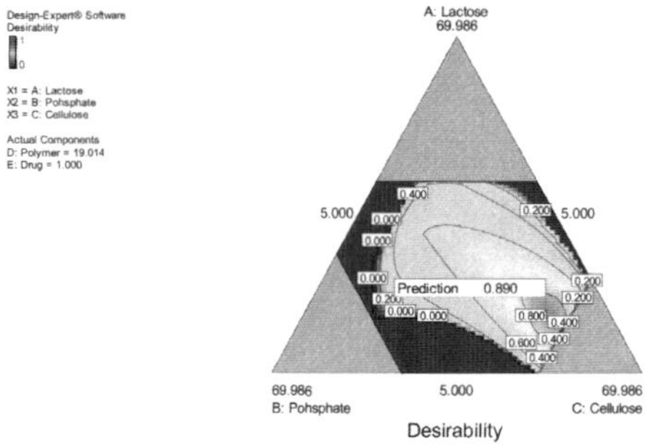

Tablet Formulation
Drug = 2%

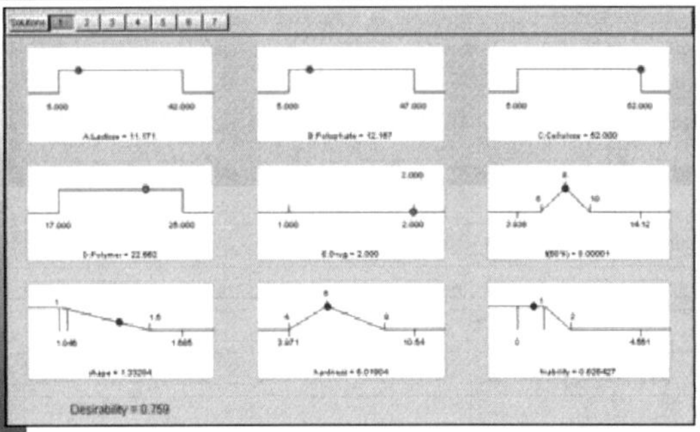

Tablet Formulation
Drug = 2%

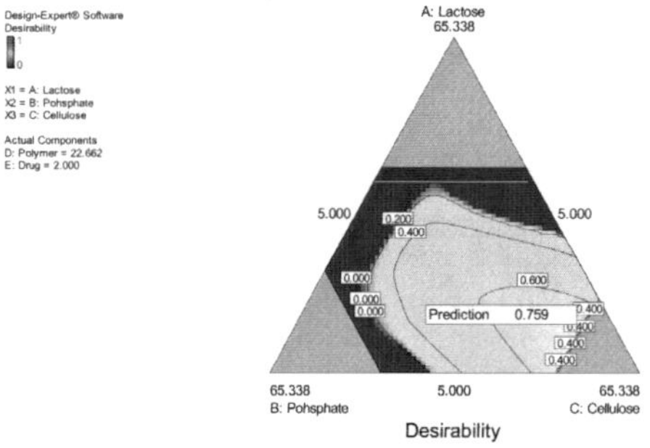

Agenda

- Mixture *(hydrophilic tablet)*
  - Build, run and analyze
  - Numeric Optimization
  - **Design Space Determination**

Design Space Determination
Quality by Design

FDA defines design space in its *Guidance for Industry Q8: Pharmaceutical Development*, as:

"The multidimensional combination and interaction of input variables (e.g., material attributes) and process parameters that have been demonstrated to provide assurance of quality. Working within the design space is not considered as a change. Movement out of the design space is considered to be a change and would normally initiate a regulatory post approval change process. Design space is proposed by the applicant and is subject to regulatory assessment and approval."

Design Space Determination
Statistically Designed Experiments

1. A region where quality product is produced.
2. Includes product formulation and/or process parameters.
3. Arrived at by iterative application of risk assessment and experimental design to knowledge space.
4. Includes evaluation of scale and equipment.

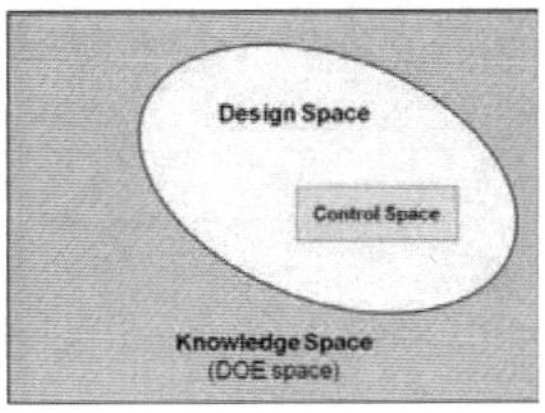

Tablet Formulation
Design Space (Drug = 1%)

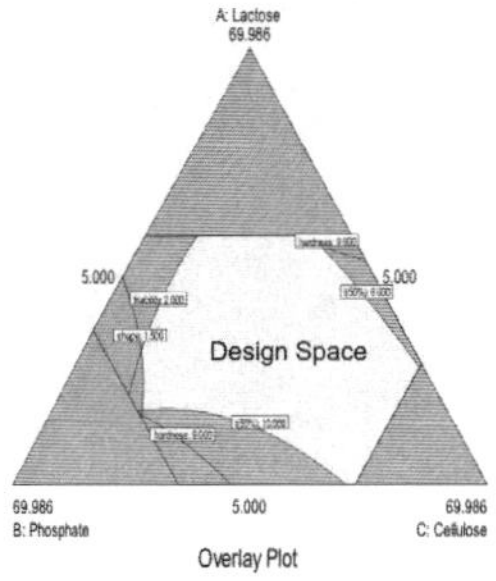

Design Space
At the Limits of the Design Space

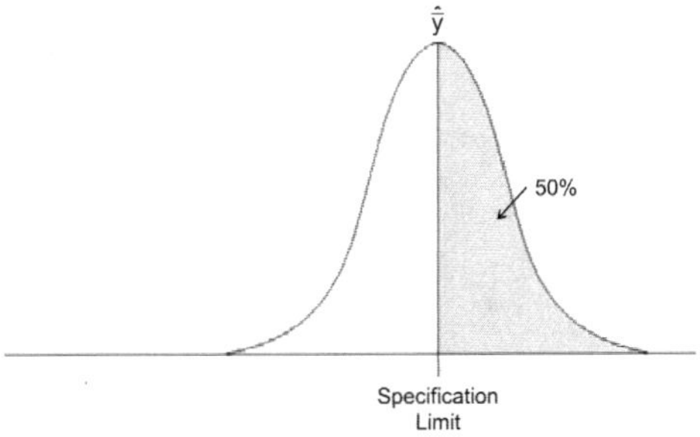

Design Space
95% One-Sided Confidence Interval

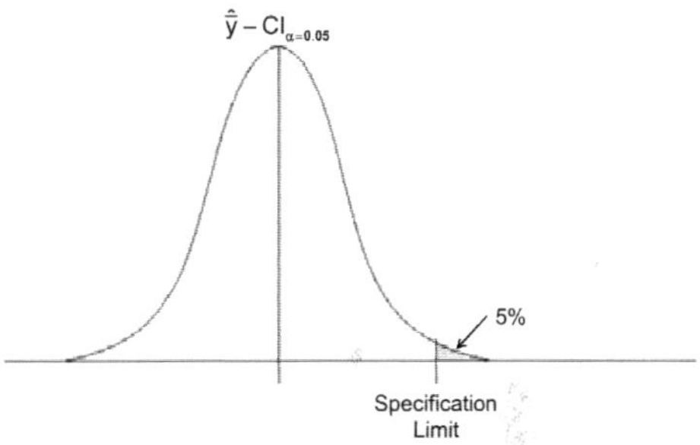

Tablet Formulation
Confidence Intervals (Drug = 1%)

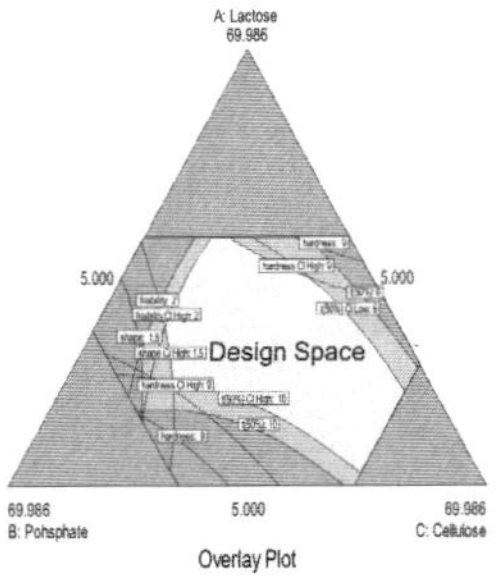

Tablet Formulation
Design Space (Drug = 2%)

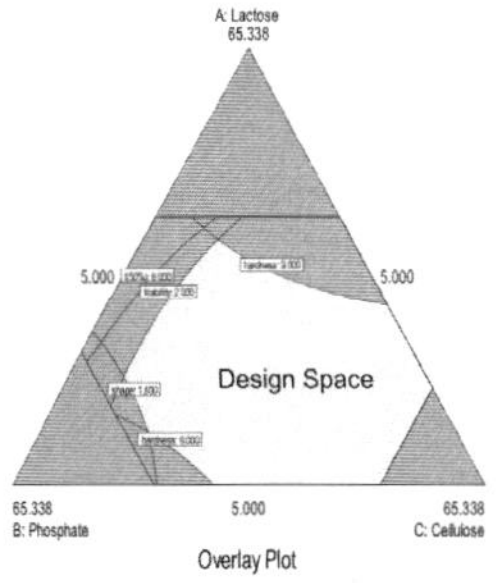

Tablet Formulation
Confidence Intervals (Drug = 2%)

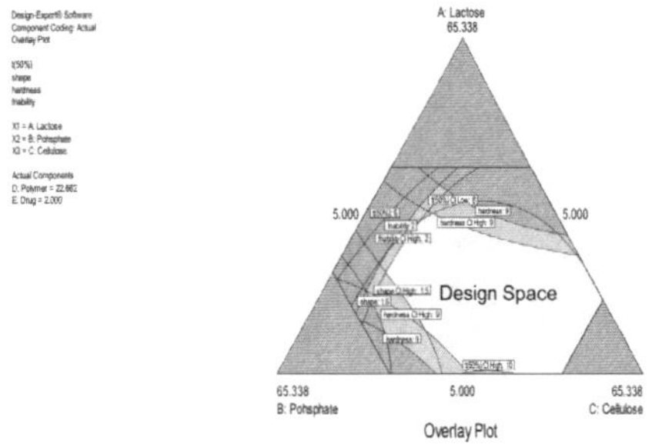

Summary

Design Space Determination via Mixture Design:

- Identify product opportunity.
- Use mixture design as an efficient method to relate critical quality attributes to the formulation.
- Use numerical optimization to find the “best” formulation.
- Use specifications and confidence intervals to define a design space.

References

1. Wendell F. Smith (2005), *Experimental Design for Formulation*, ASA-SIAM Series on Statistics and Applied Probability.
2. John A. Cornell (2002), *Experiments with Mixtures*, 3rd edition, John Wiley.
3. Mark Anderson and Pat Whitcomb (2005), *RSM Simplified*, Productivity Press.
4. Moheb M. Nasr, Ph.D., CDER, FDA, FDA Quality Initiatives Workshop, February 28, 2007, North Bethesda, Maryland.

Thank you for your attention!

Mixture Design and Design Space

Pat Whitcomb
Stat-Ease, Inc.
612.746.2036
fax 612.746.2056
pat@statease.com

Mark Anderson
Stat-Ease, Inc.

2010 AIChE Annual Meeting

(697e) A Framework for in-Silico Formulation Design and Optimization Using Multivariate Latent Variable Regression Methods

Salvador Garcia-Munoz, Process Modeling & Engineering Technology, Pfizer Global Research & Development, Groton, CT and Mark Polizzi, Pharm. Dev., Pfizer Global Research & Development, Groton, CT

Quality by Design is a technical exercise that should be performed upstream in the design of the production train for a given product. Although each processing step will have its own inputs and outputs that effect product quality, it is clear that raw materials properties are important for the development of any process. This work addresses the first step in the propagation of raw material effects by estimating the properties of the initial blend of powders for an oral dosage form.

Many methods for mixture models and predictions can be found in literature. However, these are typically "black-box" approaches (e.g. Neural Networks) which are non-interpretable as they provide no fundamental understanding of the relationships between properties of the raw material and the end product quality attributes (Takayama et al. 2003). These fitted models are also known to be unstable upon inversion (Turner & Guiver 2005). These limitations can be overcome with the use of a multivariate latent variable method. The parameters of the fitted models provide an understanding of the mechanisms through which the predictions are made allowing the validation of the model from its adherence to fundamental laws of physics and chemistry rather than validating the model solely on its statistical properties. Also, these models can be readily inverted (Jaeckle & MacGregor 1998) which empowers the use of the model within an optimization framework where blends can be designed based on a set of desired formulation properties.

The use of multivariate regression models to handle properties of mixtures has been widely addressed in literature. The most recent work (Muteki & MacGregor 2007) proposes the use of the LPLS model to handle such data structures. Although the LPLS is currently the only method that will provide loading coefficients to interpret the effect of raw material properties on the product quality, it does not provide the ability to use the data on a completely new material that has not been used before in a formulation. This work addresses this limitation by introducing a new method where the properties of the raw materials are reduced by PCA before forming the L-shaped data block to be analyzed. The predictive performance of the proposed method is shown in contrast to the LPLS and the ideal mixture approach using data collected throughout 10 years of oral dosage development. After model cross-validation, the framework was further validated by contrasting the prediction results against data acquired in the laboratory. Performance of the blend in multiple compact mechanical, and powder attributes was correctly classified within previously established performance categories.

This predictive tool has been used for pre-screening of potential formulations that lie within the range of the data used to fit the models. It has the potential to reduce the number of experiments needed for process development by pre-selection formulations which have desirable properties. The use of the framework is illustrated in estimating blend properties of a new API. Due to the high dose requirement (600 mgA) the formulation needed to contain the highest API loading possible to produce a tablet of an acceptable size. The proposed model was used to predict blend properties at several different API loadings to determine the upper limit of API loading by where the blend material no longer exhibited acceptable mechanical and flow characteristics.

References

Jaeckle, C. M. & MacGregor, J. F. 1998, "Product Design Through Multivariate Statistical Analysis of Process Data", AICHE Journal, vol. 44, no. 5, pp. 1105-1118. Muteki, K. & MacGregor, J. F. 2007, "Multi-block PLS modeling for L-shape data structures with applications to mixture modeling", Chemometrics and Intelligent Laboratory Systems, vol. 85, pp. 186-194. Takayama, K., Fujikawa, M., Obata, Y., & Morishita, M. 2003, "Neural network based optimization of drug formulations", Advanced Drug Delivery Reviews, vol. 55, no. 9, pp. 1217-1231. Turner, P. & Guiver, J. 2005, "Introducing the bounded derivative network--superceding the application of neural networks in control", Journal of Process Control, vol. 15, no. 4, pp. 407-415.

See more of this Session: Model Based QbD in Drug Substance or Drug Product
See more of this Group/Topical: Topical i : Comprehensive Quality by Design in Pharmaceutical Development and Manufacture

(697f) Modelling and Informatics Challenges IN Film-BASED Drug Formulations and Manufacture

Maria Elisa Luque[1], Bo Zhou[2], Rodolfo Pinal[2], G. V. Reklaitis[1] and Venkat Venkatasubramanian[1], (1) Chemical Engineering, Purdue University, West Lafayette, IN, (2)Industrial and Physical Pharmacy, School of Pharmacy, Purdue University, West Lafayette, IN

Drug-loaded thin films are a novel pharmaceutical administration form that is creating an increasing interest in the pharmaceutical industry due to its versatility of application and the feasibility of continuous manufacture. However, accomplishing this in industrial scale requires addressing the modeling and informatics challenges that arise in going from concept to reality. One has to deal with wide ranging data, information and models from pre-formulation studies, product formulation, process development, and manufacturing. We plan to address these challenges by first designing an appropriate ontological informatics framework for describing the whole film casting manufacture process.

The film manufacture process consists of three main steps:

(1) Synthesis of sub-micron engineered particles (2) Particle-loaded polymeric suspension formulation and preparation (3) Thin-film casting

Particle-loaded polymeric suspension preparation and thin film casting are assessed employing a response surface experimental design. The study correlated formulation parameters like particle load, solvent composition or plasticizer amount and processing parameters like drying temperature or wet film thickness over a range of properties interesting from a processing and a formulation point of view. The statistical results obtained from this experimental design are useful for:

(1) Gaining a deeper understanding in the effect of processing conditions in drug loaded film manufacture (2) Provide formulators with tools that will enable them to formulate films with the desired final properties (3) Obtain simple descriptive models that could be applied later on process control.

With the information obtained from experiments and process description, an appropriate ontological informatics framework was designed by exploiting our prior contributions in this area, namely, the Purdue Ontology for Pharmaceutical Engineering (POPE). We will report on the details of this framework, the integration of diverse sources of information ranging from process control applications to product characterization or film formulation models and guidelines, as well as lessons learnt from the implementation of the experimental testbed in drug loaded films manufacture

See more of this Session: Model Based QbD in Drug Substance or Drug Product
See more of this Group/Topical: Topical i : Comprehensive Quality by Design in Pharmaceutical

(697g) Quality by Design Approach to Pharmacokinetic Modeling. A Case Study On Cyclosporin A

Eric Lueshen, Cierra Hall, Andrej Moša• and Andreas A. Linninger, Bioengineering, University of Illinois at Chicago, Chicago, IL

Drug dose response experiments in animals provide invaluable information about the pharmacokinetics and pharmacodynamics of new therapeutic agents. The drug response curves in various organs such as the liver, heart or brain hold quantitative information to ascertain the laws of drug fate in the living organism. Yet, rigorous analyses of drug response curves are not widely practiced in the pharmaceutical industry.

Therefore, this article proposes an automated methodology for the rigorous assessment of a multivariate physiologically-based pharmacokinetic model (PBPK) in an organism such as a rat. The application concerns preclinical stages of new drug design when therapeutic doses need to be determined. Currently, dose estimation is typically performed on animal species during preclinical trials; interspecies extrapolation is then done with large uncertainties about the global transport mechanisms, the clearances and mass balance errors. Our proposed methods use a detailed vasculature network to compute steady state blood perfusion through each organ based on experimental flow resistances. Allometric scaling laws allow to account for differences among individuals of a species based on weight. In addition, changes related to age or pathological conditions such as obesity or diabetes can be accounted for. Quantitative interspecies scaling can also be easily modeled as the system is based on first principles equations and is independent of the PBPK model choice. For parameter estimation, a set of experimentally measured concentration profiles is used as input into each subsystem, such as blood, plasma and tissue. A case study on Cyclosporin A delivered by a bolus injection will be presented. The solution for the estimation problem using mathematical programming techniques yields the desired set of kinetic parameters together with the best-fit drug concentration profiles over time. This method also allows for automatic evaluation of model quality based on the least squares minimization to the experimental data set. Different PBPK models can be compared, thus helping in selection of proper kinetics. Clinical studies would then be viewed as confirmatory performance testing of the model`s prediction.

In conclusion, this paper presents a method for semi-automated physiologically-based pharmacokinetics model selection based on experimental datasets of drug delivery. The results include time-dependent drug concentrations in each organ and provide valuable insights into the mechanics of drug transport in a study organism. The successful application of the proposed methods will lead to better design of preclinical trials, more knowledge gain from preclinical animal experimentation and eventually lead to shorter drug development times.

AUTHOR INDEX

AUTHOR INDEX

American Institute of Chemical Engineers
3 Park Avenue
New York, New York 10016-5991

CURRAN ASSOCIATES INC
proceedings.com